BATTLE

The Nature and Consequences of Civil War Combat

EDITED BY KENT GRAMM

THE UNIVERSITY OF ALABAMA PRESS
Tuscaloosa

Copyright © 2008
The University of Alabama Press
Tuscaloosa, Alabama 35487-0380
All rights reserved
Manufactured in the United States of America

Typeface: Caslon

∞

The paper on which this book is printed meets the minimum requirements of American National Standard for Information Sciences-Permanence of Paper for Printed Library Materials, ANSI Z39.48-1984.

Library of Congress Cataloging-in-Publication Data

Battle : the nature and consequences of Civil War combat / edited by Kent Gramm.
    p. cm.
    Includes bibliographical references.
    ISBN 978-0-8173-1622-8 (cloth : alk. paper) — ISBN 978-0-8173-8055-7 (electronic) 1. United States. Army—Military life—History—19th century. 2. Confederate States of America. Army—Military life. 3. Battles—United States—History—19th century. 4. Combat—History—19th century. 5. Military art and science—United States—History—19th century. 6. Soldiers—United States—History—19th century. 7. Soldiers—Confederate States of America—History—19th century. 8. United States—History—Civil War, 1861–1865—Medical care. 9. United States—History—Civil War, 1861–1865—Social aspects. 10. United States—History—Civil War, 1861–1865—Psychological aspects. I. Gramm, Kent.
    E607.B38 2008
    973.7—dc22

2007045205

Contents

Editor's Introduction

Kent Gramm

One of the most harmful consequences of the Civil War results from our very interest in the war, and our attraction to it. As a Civil War buff, you can feel your own springtime in the jaunty early days of the Southern Confederacy; you can march into an everlasting blue-sky summer with the ragged young veterans of the Army of Northern Virginia; you know the meaning of hope as, book after book, you win a brilliant victory under Robert E. Lee and Stonewall Jackson, and follow the Plumed Knight, Jeb Stuart, on a gallant dash toward the Potomac River. At Gettysburg, you feel the glory of the tragic hero, battling unyielding and unvanquished against the sure and unforgiving Fates. As William Faulkner wrote, for every Southern boy it can always be the early afternoon of July 3, 1863, and forever victory can be just an hour away. The battlefields are beautiful now. The trees bloom in spring and die in brilliant reds and yellows in autumn, and in between it is a timeless summer in which you can feel the years slip away and it is the 1860s again. You are young again yourself, for you have your own history of visiting these timeless fields.

You have learned from the Iron Brigade what it is to be a hero, standing and dying in front of Seminary Ridge at Gettysburg, so "that nation" of the unfathomable Abraham Lincoln can have a new birth of freedom. You revisit the grim partings back home in Ohio, Pennsylvania, New Hampshire, and Minnesota as young men, determined to save freedom for unknown immigrants and for unborn generations, leave their wives and sweethearts and mothers to march under the Stars and Stripes, *for us,* into "the awful shock and rage of combat." We sing again their songs, and we rally 'round the flag. For the Northerner as well as for the Southerner, it can always be early afternoon, July 3, 1863, when the hopes of a nation are still young and still full, and a kind of clarity and innocence are still poised to win the future—and the smoke and noise and dirt of the twentieth and twenty-first centuries have not yet swept in behind the buzzing machines of our age.

Who would not love such a war? But that was not the Civil War as its soldiers fought it. That is a war of fantasy, myth, and entertainment. General Sherman described the real Civil War: "War means fighting, and fighting means killing." By replacing this actual Civil War with an imaginary and beautiful war—surely a contradiction in terms—we misunderstand our own natures, and we allow ourselves to fall for what Wilfred Owen called the "old lie": that it is sweet and seemly to die for one's country. Falling for the old lie, we enter more easily into what should be entered into only as one would enter a corridor to hell: you go that way only because all the other ways are shut. War is the nastiest of human failures. It is all right if we learn to love the people and the times of the Civil War, as long as we learn also to be careful. We had better not love the war so much that we willingly repeat it, in any form.

Today we seem to be embattled again, in what John F. Kennedy referred to a generation ago as "a long twilight struggle," unsure of who our friends and enemies are, confused as to what we ought to give up and what we must preserve, ignorant of what our actions today will bring tomorrow. The lessons of the Civil War are not out of place. But the first lesson is that the Civil War was a war.

I was reminded of these things when I read *Doing Battle* by Paul Fussell. That book recounts Fussell's experience as a young officer in the Second World War and its results on him psychologically and philosophically. As Program Director of the Seminary Ridge Historical Preservation Foundation in Gettysburg, I am charged with arranging annual symposia which are meant to educate the public on topics related to the Civil War but of wider interdisciplinary and national interest. I decided to ask Professor Fussell to speak at the first Seminary Ridge Symposium as a way of fixing our attention on the essential aspect of all our conferences on the Civil War: warfare itself.

Paul Fussell is one of America's most distinguished thinkers and writers of nonfiction. His *The Great War and Modern Memory,* about the effects of World War I on the western world, won the National Book Award and the National Book Critics Circle Award. In addition to scholarly books, he has written *Wartime: Understanding and Behavior in the Second World War; Thank God for the Atom Bomb and Other Essays* and *BAD: or, The Dumbing of America; Class: A Guide through the American Status System.* He has also edited

The Oxford Book of Modern War and most recently has published *Uniforms: Why We Are What We Wear*. (This is by no means an exhaustive list of Professor Fussell's publications.) His written presentation to the 2001 Symposium is reprinted with permission of Transaction Publishers; and we also print here his prefatory remarks, which riveted the attention of the audience and, more than a program director could have hoped, set the tone of the symposium. Taking place less than two months after the events of 9/11, Professor Fussell's remarks and the entire content of the Symposium were unexpectedly pertinent—and they remain so.

D. Scott Hartwig, Supervisory Historian with the National Park Service, Gettysburg National Military Park, has prepared for this volume a definitive description of what combat was like for the Civil War soldier. (A shorter version was delivered at the 2001 Symposium.) Countless amateur and professional historians have had the same experience I had in contacting Scott Hartwig for the first time about a dozen years ago. Though not knowing me from Adam, this busy historian gladly shared with me his vast and minute knowledge of the Civil War and Gettysburg in particular, and directed me to the most pertinent and helpful sources pertaining to what I was looking for. Author of *The Battle of Antietam and the Maryland Campaign* and *A Killer Angels Companion*, Mr. Hartwig has co-authored books on Pickett's Charge and the monuments at Gettysburg and has published essays and articles too numerous to list here.

The Symposium was interested not only in battle, however; it concerned itself also with the consequences of battle. The most obvious consequences were medical. Like Scott Hartwig's essay, Dr. Bruce Evans's essay on the medical effects of battle should prove to be a standard short work in its field. It deals with Civil War wounds and medical practice authoritatively, yet in language understandable to the lay person. I have had the honor and pleasure of knowing Bruce Evans since we were little fellas wondering how to pronounce such names as *Shiloh* and *Antietam*. He is now a distinguished neurologist at the Mayo Clinic, in Rochester, Minnesota. He has devoted himself to a broad and minute study of Civil War medicine, resulting in a book on the subject and in many lectures and talks. His essay provides a healthy antidote to a sugar-coated and narcotic view of warfare.

Not a very long step removed from the physical results of battle, and lasting every bit as long, are the psychological results. Eric T. Dean's controversial book, *Shook Over Hell: Vietnam, Post-Traumatic Stress Syndrome, and the Civil War*, is an examination of what can happen to individuals who experienced Civil War fighting, not only having wounds inflicted upon them but having to wound and kill other human beings. Dr. Dean's study was comparative, drawing conclusions related to the American experience in the Civil War and the Vietnam War. A practicing attorney with a PhD in history from Yale, Eric T. Dean published an earlier version of the essay included here in the periodical *War in History*, whose editors have kindly given us permission to use materials from that essay.

So much for the individual. What were some of the wider consequences of the Civil War—national consequences? How we think of a war governs its results. The Civil War has become a story of our identity—a "myth." This usage of the word "myth" is not the everyday, casual usage that equates "myth" with "falsehood." On the contrary, a national myth becomes a reality. James M. McPherson, in *This Mighty Scourge: Perspectives on the Civil War* (New York: Oxford, 2007, 94), defines such a myth as "the collective memory of a people about their past, which sustains a belief system that shapes their view of the world . . ." As a story of our identity, a myth is a story of power—power to shape who we are and what we do, and what we will become. In what way has the Civil War become a story of identity, a national "myth"? Alan T. Nolan addresses that question through the medium of the Lost Cause Myth and finds that some of the salutary results of the war were to a large extent wiped out for at least a century by a self-serving and historically dishonest interpretation of the war. *The Iron Brigade*, published by Mr. Nolan a generation ago, has been listed as one of the 100 best books ever written on the Civil War—a considerable distinction, considering that roughly one hundred thousand books have been published on the war. Like Eric T. Dean, Alan Nolan was a practicing attorney until retirement: his landmark *Lee Considered* subjected the legendary general's reputation and image to intense and merciless courtroom-like scrutiny. The results were a rising wave of reevaluation of General Lee by historians—together with monumental irritation and everlasting condemnation on the part of devotees of Robert E. Lee and the Southern Cause in general. As a comfort to them, Mr. Nolan announced at the Symposium that he would withdraw his candidacy for Sheriff in Fulton County, Georgia. Mr. Nolan's chapter is transcribed from his oral presentation; it is a pleasure to be able to transmit some of the immediacy, pungency, and candor of Mr. Nolan's talk to our audience.

This editor's essay, "Numbers," attempts to use a few common numerical facts to get at the nature of Civil War combat and its consequences. How terrible was a Civil War battlefield? Did its soldiers' experience warrant comparison to contemporary war? How widespread were its medical results—and was there enough "trauma" to warrant serious individual and national "post-traumatic stress"? How good at killing were Civil War soldiers? Who made better soldiers, Southerners or Northerners? Was Grant better than Lee? What factor actually won the war? The essay attempts to assess the war's effects upon the people of the 1860s and to identify the consequences upon ourselves, the Americans of today. It is almost a commonplace to say that the Civil War made us who we are, but is it true? And if it is true, what does it mean? Exactly what is there about us that the Civil War made?

It is appropriate at this point to thank all who attend these conferences, to those of you who have purchased this volume, and especially to the writers who have graciously contributed their essays to this book. All proceeds will go to the Seminary Ridge Historical Preservation Foundation and to the Seminary Ridge Symposia. The Foundation preserves Old Dorm and other buildings on private property as priceless memorials to those who fought during the crucial first hours of the Battle of Gettysburg, and it is to be hoped that the Symposia will continue to add to our knowledge and thought regarding the Civil War and its meaning for "one nation, under God, with liberty and justice for all."

I

Reflections on the Culture of War

Paul Fussell

Some Prefatory Remarks

The description of this Symposium given by Kent Gramm is very honest. He says that what we're going to deal with is "fighting itself, and what it means for those involved." I am glad he used the word "fighting" instead of "combat." "Combat" is usually a sort of heroic euphemism. There's no combat involved. It is brutal, man-to-man fighting. And you had better win, if you expect to come home—win by any possible method, fair or foul. That is important to understand.

It may help if I indicate that every infantry battle has three main characteristics. First of all, there is *menace* of some kind. One perceives that some people are going to try to kill you in the next few minutes. It starts with menace. Menace, of course, generates *fear,* the second element of infantry fighting. Fear will ruin you and the people you are with, unless you have the third element, which is *obedience.* Obedience is the ideal, even if half your troops run away. (They may do that, to your surprise.) So, we could say that the size and the weight of each of these three elements, and their relation to each other—understanding that they are always there—represent a way of understanding infantry warfare.

Now, I have a little comment to make about infantry combat in the Civil War. The ground warfare which I knew in the 1940s was remarkably like the battles between the United States of America and the Confederate States of America in the 1860s. We used rifles, the way they did; we used artillery; and we used bayonets. (I think they used bayonets more viciously than we did. We used bayonets so they could be seen by adolescent German boys—to scare them to death before we even had to kill them. And they're very effective—absolutely stomach-turning—when you see them in the distance, gleaming in the sun. But we didn't really stick them in anybody's body. Some people did, but not the people I was with. We used them as a scare device.)

There are of course large differences between infantry behavior in the Civil War and the Second World War. For example, we no longer had swords for the officers. And we also had wicked things unknown in the relatively civilized Civil War. We had booby traps. And preeminently, we had antipersonnel mines. (They're still going off, as you know, all over the world.) These mines were very vicious because they blew your foot and ankle completely off. And increasingly, as the war went on, they were made of plastic, to defy electronic mine detectors. We also of course had the machine gun and machine pistol, unknown in the Civil War—at least as sophisticated instruments, the way ours were.

The infantry was said to travel in trucks. People talk about the Second World War, people who know nothing about it. People say it was a "war of mobility." To that I would utter a very obscene answer. It was *not* a "war of mobility" for most of us. The infantry almost never traveled in trucks. We walked—or, euphemism—"marched." The Germans used horses to pull the loads that we trucked; that is, rations, ammunition, artillery pieces, and so on. I fought on what was called the Western Front for five months and I never, *never* saw an airplane. Never. It was infantry warfare, very much like Civil War stuff.

The day I was wounded in France, we had undertaken an attack—a frontal attack—on a little woods which we knew was full of the enemy. How did we do it? We formed a big skirmish line just like the line of Pickett's Charge, and we came running down the hill shouting and displaying bayonets, trying to scare those kids to death. It didn't quite work. But it was a Civil War operation. There was nothing modern about it at all. And therefore it's interesting for people like me to study the tactics of the Civil War.

I have objected seriously to some public statements about warfare. Here is one from an official report on the medical service in the European theater of operations. It's about the trench foot problem in the winter, which caused as much non-military damage to the troops as malaria did to those in the Pacific theater. Here is the statement I read that set me off: "Elaborate instructions were issued to front line units about how to avoid trench foot. Radio broadcasts carried the message to front line troops." Now from that you might—if you were innocent—get an image of the troops sitting around like people listening to Jack Benny, in a nice, comfortable house with carpets and overstuffed furniture, listening to the radio for instructions. There were no radios. We had radios, but only for tactical purposes. There was no entertainment by radio at all. (By the way, I didn't hear the famous German song "Lili Marlene" until I got back to the States a year later. I never heard it over there.) And so I wrote "Radios? Are you kidding?" There were no radios for us.

My experience has been as a junior infantry officer, performing as a rifle platoon leader, a second lieutenant. It was a very different world then, as we all know. For one thing, it was an atmosphere of a very similar kind of verbal sincerity to what you find all the way through the writings of soldiers in the Civil War: sentimental, sincere, un-ironic, un-nasty. In 1860 there was almost nothing of the sort of things we take for granted, which totally change one's use of the language. For example, there was virtually no divorce, and there was no breakup of families. The powerful pull of the big cities had not yet weakened rural life and farming. Society tended to be cohesive; society was hierarchical. Now, I am sure some of you know much more than I do about the relationships between the officers and the men on both sides in the Civil War. From what I've read, they tended to respect their officers. That would strike us as preposterous in the Second World War. We hated our officers. As a junior officer, twenty years old, I knew that I was despised by kids that I was leading who were twenty-one and older, and who were better than I was at it.

The Culture of War

My friends and I sometimes play a game that you might enjoy; we call it "oxymoron." The object is to come up with phrases that, while superficially plausible, prove on skeptical examination to involve intellectually comical contradictions in terms. Take, for example, creation science, or journalistic ethics, or the Maoist concept of a cultural revolution. How about the term scholar-athlete or, looking toward the university faculty instead of the students, the scholar-activist. That is actually a phrase the *Washington Post* used recently to describe the newly appointed president of the University of the District of Columbia. Some deeply cynical player of the game "oxymoron," contemplating much of higher education today, might go so far as to propose as the winning oxymoron: college education.

Now I start this way because my title, "The Culture of War," might be regarded as that kind of flagrant oxymoron.[1] And so it would be if I were evoking the term "culture" in any artistic or intellectual sense, implying within the armed forces a considerable amount of viola playing, classical acting, drawing, painting, poetry, fiction writing, and difficult reading. But actually it is not these sorts of things that I am trying to suggest by the word "culture." I am using it in a quasi-anthropological sense: the way T. S. Eliot used it when he wrote a book called *Notes Towards the Definition of Culture,*[2] a book in which he considered the possibility of a healthy and interesting society based on something like religious principles. In that book, Eliot understands how much, as he puts it, is embraced by the word "culture," a term not designating merely artistic or ennobling activities but the general forms and usages and techniques of a given society, including military society. To Eliot, culture includes all the characteristic activities and interests of a people. Using the British people and their culture as his examples, he goes on to list as components of British culture these things: Derby Day, the Henley regatta, dog racing, dart boards, boiled cabbage cut into sections (a rather disgusting idea surely), nineteenth-century Gothic churches, and the music of Sir Edward Elgar. In the same way, probably any one of us could make a list of things comprising the culture of war. At the outset I should warn you that the items I mention are collected not by any military strategist, theoretician, historian, or scholar. They are the views of a superannuated, badly wounded, former infantry lieutenant, a one-time rifle platoon leader who fought in World War II in Europe and commanded forty terrified young Americans, many of whom were killed or cruelly wounded. Thus, if the word "culture" presents some problems, the word "war" will present even more.

The truth is that very few people know anything about war. In an infantry division, for example, fewer than half of the troops actually fight, that is, fight with rifles, mortars, machine guns, grenades, and trench knives. The others, thousands upon thousands of them, are occupied with truck driving, photocopying, cooking and baking, ammunition and ration supplying, and similar housekeeping tasks. Now those things are no doubt necessary, but they are hardly bellicose; they don't provide the sort of experience required to define what the word "war" might mean. This is the reason why most combat veterans tend to smile cynically and sardonically at veterans' reunions when those reunions are attended by very large numbers. Very few of those attending, the real veterans know, deserve to be there. For most soldiers participating in World War II, the war meant inconvenience rising sometimes to hardship, enforced travel and residence abroad, unappetizing food, and the absence of tablecloths or bedsheets. For those unlucky enough to be in the forward combat units, the war meant death or maiming, usually in extraordinarily dirty and undignified circumstances. At the very least, for most it meant a rapid and shocking metamorphosis from boyhood innocence to adult cynicism and bitterness. It is an experience remembered so vividly even at this distance that it has inducted me into my understanding of the culture of war. It is a culture hard for civilians to understand, because civilians occupy a world, thank God, that is in large part rational and predictable, a world that makes sense in an old-fashioned way.

Now let me illustrate what I mean. A while ago I was telephoned by a lawyer in New York City. He indicated that he was conducting a course for cadets at West Point, a course in the relation between language and violence. This course focused on the deformations of language required for the registration of non-rational violent behavior. He asked me to take part in a class on this topic and I agreed cheerfully. He then specified the subject further. He was going to focus, he said, on the after-action reports from combat units, and he wanted me to indicate what problems I had experienced in writing my after-action reports. What problems had I had adapting normal language to this special use? For example, what euphemisms, in any, were employed in these after-action reports? What were the temptations that I felt to provide rational motives for violent or inexplicable events?

As this phone call went on, I confess that I suffered an outburst of extreme anger, the sort of thing that is common among infantrymen reinstalled in an optimistic and unimaginative civilian culture. With some passion, I asked this lawyer, have you ever been in combat? He answered no. I then explained, with elaborate sarcasm, that I never heard of such things as after-action reports from small assault units. Perhaps they had some existence at battalion or regimental level, but not down where the fighting was. How, after all, could one pull oneself together to compose an after-action report with pencil and paper when you had the following after-action features to attend to: First, the question of what to do with the six German prisoners the assault had just yielded. How to keep a very angry private who had seen his buddy's eye shot out from doing what he really much wanted to do, to kill all the prisoners? Second, after-action you had to clean up the mess. This meant taking care of the wounded, some of whom are suffering intense, unrelievable pain because the morphine is already exhausted. Third, after-action you had to reposition your soldiers to repel a German counterattack, and you had to jolly them up to make them work to continue fighting the war in the prescribed manner. Fighting the war after-action is going to be very difficult because your sergeant is over there crying. Fourth, how could a junior officer, like me, write an after-action report when his hand was covered with the blood of one of his men whose wound in the back I had ineffectively tried to bandage while the bullets and shell fragments were flying around? Fifth, given all this, how could such a person have waited a day or so to file his after-action report in a calmer mood, when a third of the men whose testimony would be required were gone, killed, or wounded? By that time he would be engaged in further violence himself. The point is that producing after-action reports is the privilege of leaders who are noncombatants, and are useful only in works of fiction.

My point is not that we did not write such reports; rather, my point is that the lawyer, a very representative human being, suffered from an extreme naiveté about the facts of war. One would expect a lawyer, in New York City especially, to be quite sophisticated about the facts of life, but here is one who imagined that the conduct of combat was rational. He was a victim of what I call "inappropriate rationalism" mixed in with a bit of inappropriate optimism as well. Those who find it hard to understand how often soldiers kill their own comrades during friendly fire episodes are victims of the same intellectual and emotional error. The culture of war, in short, is not like the culture of ordinary peacetime life. It is a culture dominated by fear, blood, and sadism, by irrational actions and preposterous (and often ironic) results. It has more relation to science fiction or absurdist theater than to actual life, and that makes it hard to describe. If you like you can regard what I have said about this bizarre and ignorant concern with after-action reports as just another bureaucratic intrusion into a place where such intrusion is entirely inappropriate and, even worse, stupid. It is especially unfortunate because it simply underlines the unpleasant fact of the military class system. On the one hand, there are the remote and privileged staffs and administrators; on the other hand, there are the troops, mostly sad conscripts, who must do the dangerous work.

The distance between serious survivors of war and optimistic onlookers can be measured by a current controversy in Britain between veterans of World War II and the government. The veterans want D-day commemorated with solemnity and sorrow. After all it marked the beginning of a battle in which 37,000 people, most of them pathetically young, were killed. The government, desirous of tourist dollars, takes a different approach. It proposes not a commemoration but a celebration, involving street parties, dances, huge reenactments, band concerts, Glenn Miller impersonators, and the like. Well, the quarrel is between those who know the culture of war, and those who think they know it, or who are prepared to profit from a misrepresentation of it. Between these two groups a reconciliation is hardly possible. A spokesman for the veterans has said that the event is being trivialized. Those who actually took part feel it was just a battle, albeit a successful one. Many of their comrades lost their lives in the process and many women were widowed. That the allies won World War II does not oblige us to be cheerful about it. Wars are won by distinction in the techniques of mass murder, and that is hardly something for people pretending civilization to be proud of. Tolstoy's words are worth recalling: war, he said, "is not a polite recreation, but the vilest thing in life, and we ought to understand that and not play at war." It will be many years—perhaps decades—before it becomes clear whether the Cold War was really necessary, or was a gratifying and profitable playing at war whose beneficiaries were not the people, but only the makers of armaments designed to become rapidly obsolete and quickly replaceable. If focusing the economy on armaments bankrupted the Soviet Union, think what it did to the United States.

Thus, while the culture of war solidifies the connectedness of those who fight, it alienates them from those who do not. It has other regrettable aspects, one of which is censorship. War kills people; the culture of war does not, but the culture of war kills something precious and indispensable in a civilized society: freedom of utterance, freedom of curiosity, freedom of knowledge. Recently, an official of the Pentagon explained why the military had censored some TV footage depicting Iraqi soldiers cut in half by automatic fire from U.S. helicopters. He explained, "If we let people see that kind of thing there would never again be any war." Now I got that quotation from a comical gift book titled *The Seven Hundred and Seventy-six Stupidest Things Ever Said*.[3] But that remark is far from stupid; it is very true and its implications spread very far. It is obvious that censorship of that type is a necessity in any modern war. It is usually rationalized by the need to keep the enemy in the dark about our plans; it is also valuable to conceal military blunders and war crimes from a public that, in the absence of censorship, might learn to be critical of the military's actions.

Now my point is simple: if you are trained to be uncritical of the military, you can easily go a little further and learn to be uncritical of government and authority, and even to be uncritical of all established and received institutions. The ultimate result is the death of the mind, the transformation of higher learning and independent scholarship into a cheering section for whatever popular notions and superstitions prevail at the moment. During wartime, and during the Cold War, we all had to pretend that the military was a force for some kind of social good. I wonder if the habit of unthinking obedience is a good one to instill in young Americans. For one thing, what is clear about the culture of war is that it is necessarily an obedience culture. In armies, as one critic has noticed, where there must be unquestioning obedience, there must necessarily be passive injustice. And not just that—the obedience culture is certain over the long run to shrivel originality and to constrict thought, to encourage witless adaptation and social dishonesty.

The culture of war is the only culture where the concept of morale is crucial, and that is a significant point. Morale is crucial in the culture of war because at all times the troops are engaging in activities sure to undermine cheerfulness and hope. They are either being bored picking up cigarette butts, or they are being dehumanized by killing their fellow creatures who, like them, are for the most part helpless conscripts who have done nothing for which they deserve to be blown to bits. In a wartime culture, censorship has the assistance of general euphemism and programmatically inaccurate language. Before long we are calling war "peace-keeping." What used to be designated aerial bombing has been euphemized into air strikes and even surgical strikes, dishonestly implying a degree of accuracy that would make combat veterans laugh out loud. Originally, artillery or mortar shells fired by mistake at our own troops were called terrible mistakes, or tragic errors. Then the euphemism of "incontinent ordinance" was devised, and finally some Pentagon genius hit upon the warmer and cozier term "friendly fire."

During the Gulf War, friendly fire caused a large share of the American casualties. Twenty-three percent of the American dead died from friendly fire. Fifteen percent of the American wounded were wounded by friendly fire. Of course, blunders are the very essence of war, which is why the culture of war is so far removed from the culture of predictability and ratio-nality. Soldiers know that mistakes are the essence of war, because they know what is likely to happen when you arm a lot of frightened boys with deadly weapons. But the public must not be told, lest their simple faith in military authority and rationality be shaken.

Transforming the ugly and shocking into the noble and bright is the business of the most popularly illustrated history of World War II. I am referring to the Time-Life volumes with titles like the *Italian War* or *Across the Rhine*. In those volumes, clear and noble cause and purpose are as-signed systematically to events which are really accidental or which are embarrassingly demeaning. Readers of those books are insulted by be-ing presumed to be incapable of confronting the truth. Everything must be transformed into fairy tales of heroism, success, and nobility. The en-tire series of books attempts to portray catastrophic occurrences in an orderly, wholesome, and optimistic fashion. For example, the shooting down of hundreds of American paratroopers during the invasion of Sicily by frightened and undisciplined American sailors, who were convinced that the large airplanes flying overhead held enemy troops, is presented in a fashion that does not show the complete bungling that occurred. The presentation of war by such dishonest means is a fine way, actually, to en-courage a moralistic, nationalistic, and bellicose international politics.

It is customary to maintain that American wars are all fought on behalf of freedom, but few notice that for the sake of freedom millions of young men are enslaved for years, shanghaied by conscription into a life whose every dimension is at odds with the idea of freedom. Flags, uniforms, bugle calls, band music, and all the trappings of military glory hardly suf-fice to persuade the hapless conscript that he is involved in the defense of freedom, especially when his weekend pass has just been canceled at the last minute in retribution for a heartfelt satiric remark that his ser-geant has just overheard. To invoke a rude term that I hope will not of-fend the reader, the culture of war is hardly separable from the culture of chicken shit.

During World War II, an Australian poet, John Manifold, wrote a poem entitled "Ration Party." It dramatizes the irony of slaves in uniform de-fending freedom. It adds to the irony by being a sonnet, a kind of po-etry normally associated with delicate or beautiful sentiments. Here is his poem "Ration Party":

Across the mud the line drags on and on;
Tread slithers, foothold fails, all ardors vanish,
Rain falls; the barking N.C.O.'s admonish
The universe more than the lagging man.

Something like an infinity of men
Plods up the slope; the file will never finish,
For all their toil serves only to replenish
Stores for tomorrow's labors to begin.

Absurd to think that Liberty, the splendid
Nude of our dreams, the intercessory saint
For us to judgment, needs to be defended
By sick fatigue-men brimming with complaint
And misery, who bear till all is ended
Every imaginable pattern of constraint.[4]

Now, the final thing I want to point out about the culture of war is that
it is necessarily adversarial and dualistic. We are here, the enemy is over
there, and a no man's land, either literal (geography) or figurative (ideology), divides us. The divisiveness at home occasioned by the Vietnam
War is an example. The divisiveness almost ruined the United States. You
remember how it went—if you opposed the war you were dishonoring
the flag and were practically a traitor. If you favored the war you were a
true American. You had to be either a dove or a hawk—take your choice.
There was no room for compromise, conciliation, or even very subtle discussion. If you were not for the war you must be for communism. It was
that attitude that finally brought down the Nixon White House.

Earlier in our history, invasion or physical pressure against American
territory were provocations leading to war. During the Nixon era, the U.S.
became "Kissingerized." No longer requiring threats to American territory, threats to American "national interest" became a sufficient reason for
sending the troops into bloody action. "National interest" is an interesting
concept because it is legally meaningless and constitutionally undefinable,
hence popular. The term "national interest" is the best gift ever awarded
to those Americans who are neurotically bellicose, but who, like Henry
Kissinger, always seem to avoid being on the frontline, preferring to serve
their country by getting others to drop bombs on people. Of course, the

people they drop bombs on, and this is notable, are always more primitive and unfortunate than themselves. They are always smaller in stature. They usually have darker skins. That is what the current culture of war seems to amount to. Clearly, we should abhor it.

Notes

1. This essay is transcribed from the author's presentation at the Mises Institute's "Costs of War" conference in Atlanta, May 20–22, 1994, and was originally published in John V. Denson, ed., *The Costs of War: America's Pyrrhic Victories* (New Brunswick, NY: Transaction Publishers, 1999). Used by permission of Transaction Publishers.

2. T. S. Eliot, *Notes Towards the Definition of Culture* (New York: Harcourt, Brace, 1949).

3. *The Seven Hundred and Seventy-six Stupidest Things Ever Said,* Ed Ross and Kathryn Petra, eds. (New York: Doubleday, 1993), 213.

4. John Manifold, "Ration Party," *Selected Verse* (New York: John Day, 1946), 72.

"It's All Smoke and Dust and Noise"

The Face of Battle at Gettysburg

D. Scott Hartwig

In his landmark study, *The Face of Battle*,[1] John Keegan called for a new approach to the study of campaigns and battles. Instead of focusing on generalship and strategy, or the detached narrative of who moved where and when, Keegan challenged historians to examine what battle was like for those who fought it. Keegan wanted to know what a soldier might have seen of the battle, how he fought, what he fought with, and why. What were the physical circumstances under which he fought? What did the battle sound like, smell like? How did the soldier control his fear? Go to his death? Keegan believed a "diversion from the rear to the front of the battlefield" was long overdue. This paper will take up Keegan's challenge. We shall move to the front, peer into the face of the battle of Gettysburg, and attempt to understand what it was like for those who fought it.

Gettysburg was the largest and bloodiest battle of the American Civil War. Nearly 165,000 combatants took part in the three-day battle between July 1 and 3, 1863, and they inflicted up to 51,000 casualties upon one another. The latter number breaks down to approximately 10,000 deaths, 27,000 wounded, and 10,000+ prisoners of war and missing. An untold number of POW's died in captivity, so the true death toll caused by the battle is not known with absolute certainty.

Because of its size, casualty total, and the fact that after the war many perceived the battle to be a major turning point in the war, Gettysburg became the most studied and written about battle in American history. Through this massive amount of literature we have learned a great deal about the generalship of the battle and about how the battle was fought, and where everyone moved to and from, but considerably less about the various experiences of battle by those individuals at the sharp end.

The Armies at Gettysburg

The battle was fought by the two major field armies in the eastern theater of the war: the Union Army of the Potomac, commanded by Major General George G. Meade, and the Confederate Army of Northern Virginia, commanded by General Robert E. Lee. These were both veteran armies, and the vast majority of their men were volunteers. Conscription had not yet started in the North, so the Army of the Potomac was composed entirely of volunteers, except for a small number of U.S. Regular Army soldiers. The Confederacy had implemented conscription in the spring of 1862 and there were conscripts in the ranks of the Army of Northern Virginia, but they represented a relatively small percentage of the army's strength, and conscripts in the Confederate army were absorbed directly into veteran units.

All of the troops in both armies were well-trained soldiers by Civil War standards, meaning they had passed through a substantial period (several months) of drill and discipline, and some time in the field, although not necessarily in combat. The majority of troops in both armies had participated in one or more major battles between 1861 and 1863, and an untold number of skirmishes and other minor encounters. With few exceptions these were seasoned troops, well acquainted with the realities of the battlefield.

The typical foot soldier, line officer, trooper, or artilleryman identified with his regiment or battery, the standard building blocks of a Civil War army. An infantry or cavalry regiment generally consisted of ten companies, which at full strength consisted of 100 men each, so a full-strength regiment numbered 1,000 officers and men. No unit in the field ever maintained full strength and the average Union regiment at Gettysburg numbered about 300 effectives while the average Confederate regiment counted about 320. A battery in the Union army contained six cannons, plus limbers and caissons, and between 100 and 130 officers and men. A Confederate battery consisted of four cannons, plus limbers and caissons, and between 65 and 80 men.

The Soldier's Perspective: "Its All Smoke and Dust and Noise"

If you asked a typical veteran what he saw of the great battle you probably would not learn very much about the battle as a whole. Ambrose

Bierce, who served in the western theater of the war, wrote: "The civilian reader must not suppose when he reads accounts of military operations in which relative position of the forces are defined, as in the foregoing passages, that these were matters of general knowledge to those engaged . . . It is seldom, indeed, that a subordinate officer knows anything about the disposition of the enemy's forces—except that it is unamiable—or precisely whom he is fighting. As to the rank and file, they can know nothing more of the matter than the arms they carry."[2] The vast majority of soldiers observed only the narrowest slice of the engagement. Many factors affected what a soldier could see of the battle. Abner Small, the adjutant of the 16th Maine Infantry, described some of the factors that limited one's observation:

> Any member of a regiment, officer or private, can have but little knowledge of movement outside of his immediate command. In all engagements with the enemy he has his special duty to perform, and no time to look with a critical eye upon his commander's conduct; he has all he can do to obey orders, and keep from running—many failed even in this . . . The idea that a soldier, whose duty it is to remain in the ranks and move in geometrical lines, has an opportunity to view a Gettysburg as he would a panorama, is absurd. . . The inequalities of the ground, the wooded slopes and deep ravines, the fog, the dense smoke, and the apparent and often real confusion of troops moving in different directions under different orders, utterly preclude the possibility of a correct detailed observation of a battle of any magnitude.[3]

Colonel John Stone, commanding the 2nd Mississippi Infantry, responding to a postwar letter requesting his recollections of the fighting on July 1, wrote, "I was very much like the French Soldier of whom you sometimes told us, who never saw anything while the battle was going on except the rump of his fat file leader. In battle I rarely knew anything that occurred beyond the immediate vicinity of my own command. In battle when I commanded a company it engaged my whole attention, when a regiment, I knew little of any other command, and so on." If this was the view of the battle from a colonel mounted on horseback, we might conclude that the rifleman in the ranks saw very little. Robert Whittick, a private in the 69th Pennsylvania, one of the regiments that absorbed the shock of Pickett's Charge on July 3, offered some insight into this in his testimony during a court case regarding the placement of a monument to the 72nd Pennsylvania. After being asked a series of very specific questions regarding events of the action he participated in, and where specific units were located during the fight, Whittick at last exclaimed, "Of course, you cannot expect me to answer these questions, because I could not see everything over the battlefield. I was looking out for myself."[4]

The front line of the two armies extended for nearly four miles, spread out over rolling terrain and hills. The frontage of the armies, combined with the terrain, precluded anyone from seeing the entire battlefield at one time. Even units that were very close to each other's sometimes were unaware of one another's presence. In recalling the fierce fighting on the Stony Hill near the Wheatfield, Confederate Brigadier General Joseph B. Kershaw wrote that he had no idea where exactly the left wing (three regiments) of his brigade was, even though elements of it were only about one hundred yards away from him. The terrain contributed to this but so did the absolute confusion and chaos of battle.

George G. Meade and Robert E. Lee, who were in a position to see more of the battle than anyone as army commanders, saw none of the fighting at Culp's Hill or Cemetery Hill. Meade did not even see the final repulse of the Confederate grand assault on July 3. He was shelled out of his headquarters by the preattack bombardment and forced to relocate. By the time he made his way to Cemetery Ridge, the attack had been turned and all that remained were casualties, prisoners, and fugitives returning to Seminary Ridge. Meade's and Lee's management of the battle relied far more on reports from staff and subordinates than on personal observation.

During periods of heavy fighting, dense black powder smoke blanketed the field like fog, greatly obscuring an individual's ability to see any distance. Christopher Smith, a gunner in Battery A, 4th U.S. Artillery, recalled that within a short time of his battery returning the Confederate fire during the great bombardment, before Pickett's Charge, "the whole landscape was enveloped in such a cloud of smoke that nothing could be seen . . . It was no longer of any use attempting to train our guns on any particular spot." Union Brigadier General John Gibbon, describing the same bombardment, wrote that, "over all hung a heavy pall of smoke underneath which could be seen the rapidly moving legs of the men as they rushed to and fro between the pieces and the line of limbers, carrying forward the ammunition." Captain George Collins, a company commander in the 149th New York on Culp's Hill, wrote, "soon after the action commenced (on July 2) the smoke became so dense that the men were unable to distinguish the enemy and were governed more by hearing than sight in firing."[5]

Gut wrenching, spine-tingling fear also limited the soldier's view of the battle. When an individual is concerned about self-preservation, he is unlikely to be interested in anything more than survival. Ben Hirst, a member of the 14th Connecticut, reflected on the terror he and his comrades endured during the July 3 bombardment and how little of the bombardment he actually observed: "How we did Hug the ground expecting every moment was to be our last. And as first one of us got Hit and then another to hear their cries was Awful. And still you dare not move either hand or foot, to do so was Death. Once I ventured to look around and just then I saw one of [the] Caissons blown up, while the same moment a Rebel one was blown up."[6]

Bernard Matthews, a soldier in the 108th New York, provided about as honest a description of what the private soldier and line officer saw of the battle as can be found. He wrote: "A soldier who is in a battle can tell you about as little about the battle as anyone in the world. It's not what you think. It's all smoke and dust and noise. At first we could see the Confederates moving around and putting up their guns. At that stage there was just occasional firing by the skirmishers and not enough smoke to hide anything. But later when the volleying began you might as well have been blind. The smoke lay over everything so that you were lucky to see the man next to you. Your ears couldn't distinguish shot from shot. It was all one roar, so that the hillside shook."[7]

The Physical Background of Battle

What was the physical condition of the combatants at Gettysburg? The battle was not an isolated incident in time. Both armies had maneuvered some 120 miles across northern Virginia, Maryland, and south-central Pennsylvania for nearly three weeks before coming to grips at Gettysburg. Except for the cavalry, the artillery drivers, teamsters, and the like, everyone marched that distance. The nature of the marching—planned, measured marches, forced marches, etc.—mattered very much to the foot soldier who made them when he arrived on the battlefield. The Army of

Northern Virginia held the initiative through the opening phase of the campaign, slipping past the flank of the Army of the Potomac and swinging down the Shenandoah Valley. By the time Union army commander Joe Hooker realized the Confederate move to the Valley was a full-scale invasion, Lee and his army were well on their way to Pennsylvania. Consequently, in the first three weeks of the campaign Confederate soldiers generally experienced planned, measured marches that were fatiguing but not debilitating. The Army of the Potomac faced the urgency of catching up with the Rebels and driving them out of Pennsylvania. The marches they experienced leading to Gettysburg were outright ordeals. R. S. Robertson of the 93rd New York offered an idea of their impact on the troops in a June 28 letter to his parents: "Weak, sore and worn out after a long and weary march, I take the opportunity of sending you a few lines . . . You may imagine how little I feel like writing when you know what I have gone through for a few days past . . . This is the hardest marching on record since the war began and we are completely used up. The sides of my feet are covered with large blisters and the soles are so sore, I can scarcely bear my weight . . . and cannot get my boots on at all, my feet are swelled to such a size . . . 54 miles in two days would be an extraordinary march on the best roads, but in the mud it was more than any army did before."[8]

Sergeant Charles Bowen, a member of the 12th U.S. Regulars, wrote diary entries for June 15 and 17 offering more evidence of the severe toll forced marches took on the Federal army in the searing June heat. On June 15 he wrote, "The heat was intense all day & the dust in such clouds that we could scarcely draw breath . . . I know of 7 men that fell dead & a great part of the men were unable to keep up with their companies." On the 17th he wrote, "Men fell by the dozens from the excessive heat . . . I stood it very well until within 2 miles of camp & then my sight began to fail & my head swim & I had to drop under a shade tree."[9]

Many soldiers, then, particularly in the Army of the Potomac, arrived at Gettysburg fatigued or ill from the effects of heat exhaustion, bad food or water, or too little water to keep their bodies properly hydrated. Maintaining an adequate supply of water once the battle was joined was not easy. Some 160,000 men and thousands of animals taxed the existing water supply and troops sometimes made do with water of questionable cleanliness. James Houghton of the 4th Michigan wrote that on the early afternoon of July 2 the regiment's officers advised the troops to fill their canteens. Houghton and some of his comrades hunted about and found a ditch "that had some stagnant water in it. We poked the scum to one side with our cups then gave the water a spat to scare the bugs and wigglers to the bottom then filled our canteens."[10]

Water, or the lack of, had an influence on the outcome of at least one action at Gettysburg—the engagement between the 15th Alabama and 20th Maine. The 15th Alabama was part of Brigadier General Evander Law's Alabama brigade. On July 2 they marched some twenty miles in nine hours to Gettysburg, arriving west of town around noon. After a brief rest, the brigade joined the flank march of Longstreet's 1st Corps, covering perhaps five more miles under a hot sun with uncomfortable humidity. The brigade deployed on the far right of Longstreet's line of battle. After forming, Colonel William Oates, commanding the 15th, sent off a water detail of twenty-two men with all the canteens of the regiment. Before the detail returned, Oates received orders to advance with the rest of his brigade.

The 15th advanced nearly one mile at the double-quick (similar to a jog) under fire of Union artillery and sharpshooters. Their advance brought them to the base of Big Round Top, which Oates chose to advance up rather than leaving the Union sharpshooters in his rear. The route Oates took up Big Round Top is difficult enough with modern hiking boots and light clothing. With wool clothing, rifles, fighting equipment, heat, and humidity, it is downright arduous. "Some of my men fainted from heat, exhaustion, and thirst," Oates recalled of the ascent. Once they reached the summit of the 750-foot hill and after a brief respite, Oates was ordered to advance and attack the enemy on Little Round Top. The regiment did so, engaging in their famous combat with the 20th Maine and pressing that unit nearly to the breaking point before Oates questioned his own regiment's staying power and ordered a retreat. At precisely this time the 20th Maine launched a counterattack that swept the 15th back up the slopes of Big Round Top. During the retreat Oates fainted from heat and exhaustion and would have been captured had not two of his men picked him up and carried him to safety.[11]

How much had a lack of water contributed to Oates's failure? Oates thought the loss "of those twenty-two men [the entire water detail was captured] and lack of water contributed largely to our failure to take Little Round Top." Although Oates may have overstated the impact of his regiment's canteens on the outcome of the fight with the 20th Maine, the lack of water certainly was a significant factor in the fight for it probably sapped the endurance of Oates and others, and their ability to maintain the battle.[12]

Hunger also affected combat endurance, although not to the degree that lack of water did. In another odd reversal of typical conditions, the soldiers of the Army of Northern Virginia were relatively well supplied with food before and during the battle, for they had been successfully foraging for supplies in Pennsylvania. The Army of the Potomac, compelled to make forced marches across northern Virginia and Maryland, were more precariously supplied. Once battle was joined, many units outmarched their supply trains to reach the battlefield and were without rations for between twenty-four and forty-eight hours. Charles Haley, a private in the 17th Maine, recalled that in his regiment on July 3, "some of us had scarcely tasted of anything but water for twenty-four hour and some thirty-six hours." Ralph Sturtevant of the 13th Vermont recalled that although his regiment received no rations between July 1 and July 3, it was not until the fighting was over that "an unusual craving for food seemed to prevail among all."[13]

The heat that prevailed during much of the battle probably had more effect upon the combatants than did hunger. On July 1, a cloud-covered sky produced some rain showers and kept the temperature down to a high of seventy-six degrees. The skies began to clear on July 2 and the temperature reached a high of eighty-one degrees. Clouds and a hot sun marked July 3, the warmest of the three days, with the thermometer reaching eighty-seven degrees.[14] Uncomfortably high humidity marked all three days, and took its toll on the fighting men, particularly on July 3. On the afternoon of that day, before the opening of the Confederate preassault bombardment, Anthony McDermott in the 69th Pennsylvania on Cemetery Ridge recalled that "the sun was shining in all its glory, giving forth a heat almost stifling and not a breath of air came to cause the slightest quiver to the most delicate leaf or blade of grass." A member of Stannard's 2nd Vermont Brigade described the heat that day as "almost unbearable." John Dooley, a line officer in the 1st Virginia Infantry, and among those who participated in the Pickett-Pettigrew-Trimble July 3 attack, wrote that after the artillery bombardment ceased and the brigades rose up in preparation for the assault "there is a line of men still on the ground with their faces turned." Among them were "the men who have charged on many a battlefield but who are now helpless from the heat of the sun."

Dooley also noted that as the regiments formed up some men "were actually *fainting* from the heat and dread. They have fallen to the ground overpowered by the suffocating heat and the terrors of that hour."[15]

Two of the most oppressive and distinctive characteristics of the battle were smoke and noise. Smoke has already been considered in its effect upon what a soldier could see of the battle. Its acrid, pungent smell burned the nostrils of everyone on the firing line and added to the general confusion of combat. But while smoke oppressed the senses, reduced vision to a few yards, and added to confusion, noise literally assaulted the entire being of the soldier. It is inconceivable to most visitors to Gettysburg how loud the battle was, and we cannot begin to comprehend the experience of battle if we do not consider the impact sound had on the soldier. The noise was beyond anything most human beings, except those who have experienced modern combat, have ever heard. If we contemplate the choking, blinding smoke that covered the battlefield in times of intense combat with a furious noise so loud one can barely think, let alone attempt to communicate, we can begin to understand the conditions under which the fighting men labored. Major Samuel Roberts of the 72nd Pennsylvania, during the pinch of the fight at the Angle on July 3, stated that "it was such a tremendous racket that you couldn't tell who was shooting." In the same action, Captain Henry Abbott, the acting major of the 20th Massachusetts, received orders for his regiment to change front to the right in order to confront the Confederate breakthrough at the Angle. "The noise was such, however," reported Abbott, "that it was impossible to make any order heard." Confusion ensued and the soldiers commenced to retire. Realizing that "an example could be seen, though words could not be heard," the officers of the 20th rushed to the front "and without further formalities the regiment was hurried to the important spot." George Hillyer of the 9th Georgia Infantry, in action on July 2 in the area of Rose Woods and the Wheatfield, wrote that he could observe Union artillery on Little Round Top "belching forth volumes of smoke all along the summit," but he could not hear them because "the roar of musketry and the shouts of our men drowned every other sound." Lieutenant Frank A. Haskell, an aide to Union General John Gibbon, in attempting to describe the deafening roar of the cannonade that preceded Pickett's Charge, wrote, "Who can describe such a conflict as is raging around us! To say that it was like a summer storm, with the crash of thunder, the glare of lightning, the shrieking of the wind, and the clatter of hail-stones, would be weak."[16]

Before the combat became one vast indistinguishable roar it was possible to pick out the individual sounds made by the various type of weapons or missiles, and by the fighting men of either army. Lieutenant J. H. Moore of the 7th Tennessee recalled that when his regiment climbed the fences along the Emmitsburg Road during Pickett's Charge, the thousands of bullets fired by the Federal troops on Cemetery Ridge struck the fence rails "with the distinctiveness of large rain drops pattering on a roof." Captain Henry T. Owen also participated in the charge, with the 18th Virginia. The sound that seemed to make the greatest impression on him was that made by canister. He described it as "a sound filling the air above, below, and around us, like the blast through the top of a dry cedar or the whirring sound made by the sudden flight of a flock of quail." Lieutenant Frank Haskell wrote that even during the peak of the cannonade "a million various minor sounds engaged the ear." Artillery projectiles made the most distinctive sounds. "The projectiles shriek long and sharp,—they hiss,—they scream,—they growl,—they sputter . . . and each has its own different note, and all are discordant."[17]

The human voice added its own discordant notes to the battle. Lieutenant Abner Small, who advanced toward Ziegler's Grove at the high point of Pickett's Charge, stated he heard something "strange and terrible; a sound that came from thousands of human throats, yet was not a commingling of shouts and yells, but rather like a vast mournful roar." Writing to a friend after the battle, Captain Francis Donaldson of the 118th Pennsylvania, whose regiment had retreated from its position near the Wheatfield when the forward line of the 3rd Corps collapsed on July 2, stated they were brought to a standstill "by a yell so fierce and terrible that the very blood seemed to curdle in our veins." The "fierce and terrible" yell came from the throats of over 1,200 men of McCandless's brigade of Pennsylvania Reserves who were moving up to counterattack the Confederates. Donaldson also recalled that these 1,200 men made "a sound as if a hurricane was swept toward us." It was, he continued, "the crushing of leaves and twigs" made by McCandless's men as they came up "in mass, at the double quick, arms at the right shoulder." They passed through Donaldson's regiment "with diabolical screeches and shouts," and drove the advanced Confederate troops back to the Wheatfield.[18]

The storied "rebel yell" was heard frequently during the battle, and apparently carried better than the cheers of Federal troops. Captain John Bigelow, whose 9th Massachusetts Battery engaged the 21st Mississippi Infantry in a truly desperate battle near the Catherine Trostle farm buildings on July 2, wrote that the Mississippians "were yelling like demons," while they attacked his guns. Lieutenant Haskell, listening to the roar of battle on July 2 from Cemetery Ridge, wrote "the fire of Artillery and Infantry and the yells of Rebels fill the air with a mixture of hidious [sic] sounds." The South Carolinians of Kershaw's Brigade advanced with "shout, shriek, curse, and yell" during their attack on Stony Hill on July 2 according to a soldier in Donaldson's 118th Pennsylvania. First Sergeant Franklin Whitmore in the 17th Maine, which fought G. T. Anderson's Georgians in the Wheatfield on July 2, described the shouts of the Georgians as "fearful yells" and "that wild yell."[19]

While the battle yells could strike fear into the hearts of soldiers on the receiving end of an attack, the most demoralizing sound was what some men referred to as the "death shriek," that a mortally wounded soldier sometimes emitted before he died. Captain George Collins told of incident that occurred on Culp's Hill on the morning of July 3, and its moral effect upon those within hearing distance. The 14th Brooklyn Infantry had moved up into a support position of the Union troops occupying the line of breastworks on Culp's Hill. The front line was engaged and many Confederate bullets that had been fired high were striking the woods on the hill and ricocheting. One of these glanced downward into the chest of one of the 14th's men. "He immediately gave a piercing scream which tore the hearts of thousands about him," wrote Collins. Thankfully, Collins continued, "his companions bore him tenderly away, and all were glad when they did so lest his condition make cowards of them all."[20]

Infantry

Skirmishing and Sharpshooting

Of the various types of encounters that occurred at Gettysburg, the largest, bloodiest, and most decisive was the clash of infantry against infantry. Infantry combat can be broken down into two types: skirmishing and sharpshooting is one, and line-of-battle fighting is the other. Skirmishing inevitably preceded the main clash of infantry, like the pattering of rain that precedes the heavy downpour of a thunderstorm. It was stan-

dard tactics in both armies to cover their front with skirmishers. Typically an infantry regiment deployed one of its ten companies as skirmishers to cover its front. The Army of Northern Virginia had modified this
in some units by creating sharpshooter battalions consisting of about 120
picked men who were organized into three companies and commanded
by a field officer, usually a major. Their duty was to provide the bulk of the
skirmish and sharpshooting duties for an entire brigade. Since the picked
men were among the best shots and most aggressive men in the brigade,
these battalions were very effective.

The duty of skirmishers was to deploy to the front of their regiment or
brigade and serve as a screen that could harass and delay an attacker if the
regiment was on the defensive, or reveal the enemy if the regiment was on
the attack. Skirmishing used open-order tactics and required a higher degree of initiative from the privates and non-commissioned officers than
the line of battle did. Skirmishers were not expected to execute maneu-
vers "with the same precision as in closed ranks," nor, the tactical manual
for skirmishers declared, "is it desirable, as such exactness would materi-
ally interfere with prompt execution." All movements were to be carried
out at the quick or double-quick step, or at the run if necessary. Officers
and non-coms were also instructed to see that their men "economize their
strength, keep cool, and profit by all the advantages which the ground
may offer for cover." It was dangerous duty and disliked by many soldiers accustomed to the close mutual support of the line of battle. There
were few rules in skirmish combat, and even fewer when it came to sharpshooting.[21]

A member of the 14th Connecticut Infantry, who were positioned on
Cemetery Ridge both July 2 and July 3 and had a particularly good vantage point from which to observe both skirmishers and sharpshooters ply-
ing their trades, penned a descriptive account of the work of these individuals. On the morning of the 2nd, when the 14th first took position on
the reverse slope of Cemetery Ridge, several members of the regiment
ventured to the crest to observe the "battle of the skirmishers in front and
the work of the sharp-shooters." Much of this fighting focused around the
farm of William Bliss, which sat almost midway between Seminary and
Cemetery ridges. On July 2 the Confederates had promptly infiltrated the
farm with skirmishers and sharpshooters who, one soldier recalled, "an-
noyed our main line and made themselves particularly disagreeable when
any mounted officers came within range of their rifles." To combat them

Union skirmishers were deployed and sharpshooters stationed on Cemetery Ridge to duel with their opposites in gray. The work of the sharpshooters drew the particular interest of the men of the 14th, and another recalled, "It grew almost fascinating, we forgetting, nearly, that the game was human. One marksman had made his quarry a wounded skirmisher (one half mile away) unable to stand, who was trying, by a series of flops, to drag his body up the slope to the shelter of his own lines. The marksman fired at him for several minutes as frequently as he could load and take aim; but we confess to a feeling of relief and gladness, and we've felt it ever since when recalling the scene, when the man let up on the poor fellow and had failed to hit him."[22]

Richard S. Thompson, an officer in the 12th New Jersey, also observed the sharpshooters on Cemetery Ridge. The Union sharpshooters, members of Berdan's Sharpshooters, arrived with "very heavy, long-range telescopic rifles, with a sort of tripod rest." The Confederate sharpshooters, Thompson noticed, soon became aware "that we were using rifles that had sufficient range, but also that they were being used with remarkable precision." This made the Confederate sharpshooters more cautious and at the flash of a rifle from Cemetery Ridge they would quickly disappear then "instantly reappear, ready to try a shot or fall back again if a second rifle flashed on our line." To counter this Berdan's men split up into teams of three men. When all three men were ready and had their rifles trained on a prospective target, the number one man would fire. The other two men counted to three then fired simultaneously, each at a specific opening. The Confederates took cover when the first rifle was fired but reappeared too late to see the flash of the other two rifles and met a bullet. "Alas!" wrote Thompson, "how little we thought human life was the stake for which this game was being played."[23]

Some of the deadliest sharpshooting of the battle took place between the Federals on Cemetery Hill and Confederates in the town of Gettysburg. Major Eugene Blackford of the 5th Alabama Infantry commanded the Confederate sharpshooters who occupied the town. Given orders on the night of July 2 by his division commander to annoy the enemy on Cemetery Hill in the morning "within all my power," Blackford deployed his men in an arc around the northeast, north, and northwest sides of Cemetery Hill. His men broke passages through houses to enable them to move about safely, and took positions in the houses, behind walls, up in church steeples, anywhere that offered good cover and fields of fire.

"Many of the men were on the roofs of houses behind chimneys, when they could pick off the gunners," wrote Blackford. The major's orders "were to fire incessantly without regard to ammunition and begin as soon as my bugle sounded." When daylight broke on July 3, Blackford observed a battery of what he thought were Napoleons on what must have been the forward slope of Cemetery Hill. "One signal from my bugle and that battery was utterly destroyed," wrote Blackford. During Pickett's Charge the Federals on Cemetery Hill were forced to mass in order to be prepared to go to the support of their comrades on Cemetery Ridge. This caused them to expose themselves more fully to Blackford and his men, who took full advantage. "I fired 84 rounds with careful aim into their midst, one gun cooling while the other was in use," wrote Blackford. "Now and then the enemy's gunners would turn a gun or two on us, and give us a shot, but this was too destructive of the lives of the gunners, so it was soon stopped."[24]

Just how deadly sharpshooters could be was illustrated by Captain William J. Seymour, a staff officer in Hays's Louisiana brigade. Before daylight on July 2, Hays's Brigade was moved out of the town of Gettysburg to a position near the present Lincoln Elementary School, and only about 500 yards from East Cemetery Hill. When it became light enough to see, the Confederates discovered that Union artillery on the hill completely commanded the approaches from town and it would be impossible to form other brigades along the line Hays's troops occupied. It was also impossible to withdraw in daylight without exposing Hays's brigade to those batteries and potentially heavy losses. So Hays's men lay behind a ridge that gave them defilade all day. Union sharpshooters on East Cemetery Hill kept their line under fire all day and Seymour wrote, "It was almost certain death for a man to stand upright." Some apparently did, or exposed enough of themselves to provide a target, for Seymour noted that his brigade lost forty-five killed and wounded on July 2 from sharpshooters.[25]

Further underscoring the lethality of skirmishing and sharpshooting was the fate of Colonel Orland Smith's brigade of the 12th Corps, who lost 44 killed and 240 wounded on July 2 and July 3 from these causes alone.

Henry S. Stevens, the chaplain of the 14th Connecticut, was present on Cemetery Ridge during part of the battle. He was fascinated by the work of the skirmishers operating in the contested ground between his position and Seminary Ridge. The advance Union skirmish line was posted approximately 200 yards west of the Emmitsburg Road. A skirmish or picket reserve (which might be a reserve company, or part of a company, or sometimes the balance of an entire regiment) occupied the road, using the drainage ditches for concealment. It should be understood that the Emmitsburg Road was nearly 200 yards in advance of the main Union line, which meant that the forward skirmisher was nearly one-quarter mile in front of the main line. "Our men lay flat upon the ground by the fence, hidden and somewhat protected by the posts and lowest rails," wrote Stevens, "Nothing was visible, usually, to fire at, yet when any movement was apparent a shot or two would follow from vigilant watchers; then the rising rifle smoke would attract retaliating shots." Things always heated up whenever a fresh party of skirmishers advanced to relieve the front line. The relieving force would move as close to the front line as was possible under cover, then dash forward at a zig-zagging run until they arrived at the front line. Then the men being relieved had to make the perilous dash back to cover. "'Relieving' seemed a misnomer," observed Stevens.[26]

At one point in the battle, four companies of Stevens's regiment were ordered to advance from Cemetery Ridge to capture the Bliss Farm, and drive Confederate sharpshooters and skirmishers from their cover. Stevens described their advance:

> The detail was taken up towards the division headquarters and then down a lane one hundred and fifty yards to the Emmettsburg road, then across the road, and then into the field beyond, covered for about one hundred and fifty paces by a knoll. When the men came into view the enemy, now well read in the business and prepared for them, there was a general firing at them from all along the skirmish line and from the host of sharp-shooters in the buildings. Then the desperate character of the sortie was fully revealed, but no man could recoil though death seemed inevitable. As to advance in any kind of a formation would but furnish a better target to aim at, the order was to "go as you please," or scatter and run. Every man was put to his mettle and ran with all his might for the barn. Nearly six hundred yards were to be covered and it was soon accomplished at such speed, but several dropped on the way.[27]

One feature of Stevens and other participants' descriptions of skirmishing and sharpshooting is the lethality of the rifle. Unlike a popular misconception about Civil War battles—that the troops of both sides stood in massed formation in sight of one another—in reality it was far too dangerous to expose troops this way unless it was necessary to relieve a skirmish line or to conduct a general attack or to repel one. Otherwise, through most of the battle very few enemy soldiers were visible to soldiers of the opposing side. Everyone kept under cover as much as possible.

Skirmishing was dangerous work that tried men's courage and their nerves. Many men dreaded it more than line-of-battle fighting, although losses in skirmishing were generally lower. A lieutenant in the 126th New York Infantry, writing home after the battle, explained that he had been on the skirmish line "skirmishing with Reb. Sharpshooters," on July 4. "I came near being hit five or six times," he wrote. "I would do anything rather than skirmish with those fellows. I never want to do it again. I will charge and repel charges but don't put me in that place again." In the line of battle soldiers could draw courage from the nearness of their comrades. But the skirmisher felt the loneliness frequently experienced by twentieth-century infantryman. His position was at the extreme front of the army, closer to the enemy than anyone else. Few of his comrades were visible to him, and unlike line fighting, the enemy were concealed, except for the brief moment they exposed themselves to fire.[28]

The Line of Battle

Skirmishing and sharpshooting were merely the prelude to the critical clash at Gettysburg, or upon any Civil War battlefield: the engagement of opposing lines of battle. The infantry line of battle was a two-rank formation, with a pace or two separating the rear from the front rank. Sergeants and lieutenants were strung out behind the rear rank to act as file closers (with each front and rear rank soldier consisting of one file). Essentially, their duty was to maintain order and prevent crowding and confusion. The drill book called for the soldiers to deploy shoulder to shoulder, with elbows touching. There were four essential reasons these close order tactics were used: 1) The drill manual both armies used to train their men specified the use of close order tactics, 2) it was relatively easy to train civilian volunteers in these tactics, 3) the formation massed the firepower of relatively slow-firing muzzle-loading weapons, and 4) the close order formation facilitated easier communications and command and control. Skirmishers and sharpshooters might annoy and harass a line of battle, but they could not deliver the firepower to dislodge one or overcome it; neither could they hope to stop an attacking line of battle. Only another line of battle could do this, except in some rare instances where artillery checked an attack without infantry supports.

Although the line of battle possessed formidable firepower, it also presented a massed target to the enemy—particularly its artillery, which could cause severe damage at long ranges and also sow disorder and confusion in a formation. On the defensive, regiments in line frequently built breastworks or threw up earthworks, or sought some type of natural cover to shelter themselves from artillery and small arms fire. By the time of Gettysburg, veterans had learned that a line of battle of riflemen under cover was very difficult to overcome with a frontal attack. But to capture terrain from the enemy, or demoralize him, it was necessary to attack, and this meant exposing your infantry in these dense formations to enemy fire. Since enemy artillery could deliver accurate fire at ranges over one mile and rifle fire could be effective at ranges out to 600 yards (even though very few infantry officers permitted their men to fire at such ranges), for an infantry attack to be successful two things were necessary. First, the position to be attacked should be subjected to an artillery bombardment that would suppress enemy artillery and disorder the defenders. Second, the attacking infantry needed to commence their attack from a covered position as close to the objective as possible. Pickett's Charge, the grand Confederate assault on July 3, where some 13,000 infantry advanced over one mile of open ground, was a rare event in the course of the war. The commander of the attack, Lieutenant General James Longstreet, had objected emphatically to making the attack because of the amount of open ground the infantry would have to cross under fire to reach their objective. Most infantry attacks at Gettysburg, and in other battles of the war, originated from points much closer to the enemy.

Once an infantry formation emerged from its cover to attack the enemy, it needed to do two things somewhat at odds with one another. First, it had to preserve its formation so that when the line neared the enemy it could deliver an effective fire and, if necessary, employ shock power to overcome the defenders. But the infantry attackers also had to cover the ground over which they would be under enemy fire as rapidly as possible. Rapid movement inevitably led to disorder, which in turn reduced firepower and could weaken unit cohesion and discipline. The fastest pace the drill book permitted a line of battle to advance was the *double quick,* which was like a jog.

The nature of infantry line-of-battle fighting at Gettysburg and the tactics infantry commanders used depended upon the type of terrain. But no matter what type of terrain was encountered, skillful infantry officers always attempted to maneuver their troops to attack the flank of the enemy. This was the most vulnerable point in a defending or attacking line. If an attacker or defender could maneuver their line onto the flank of their opponent, they could deliver their full firepower along the length of the enemy line, while the enemy could only reply with a limited number of weapons at the end of their line. A unit that was "flanked" usually had little choice but to retreat and hope to find a position that offered some cover where it could be rallied and reorganized.

Although officers and file closers continually strove to preserve order in the line of battle on the attack or defense, it should not be imagined that they always succeeded. Even in units that did maintain their cohesion, their formations bore little resemblance to drill field lines when they came up against the broken terrain that marked much of the Gettysburg battlefield, and when they came under fire. Val C. Giles, a private in the 4th Texas, whose unit passed over the western slope of Big Round Top and made several efforts to dislodge Union defenders on Little Round Top, wrote that during the most intense fighting, "Confusion reigned supreme everywhere . . . Every tree, rock and stump that gave any protection from the rain of minie balls, that poured down on us, from the crest above us, were soon appropriated. . . By this time order and discipline were gone. Every fellow was his own general. Private soldiers gave commands as loud as the officers—nobody paying attention to either. To add to this confusion, our artillery on the hill in our rear was cutting its fuse too short. The shells were bursting around us, in the treetops, over our heads, all around us."[29]

Another member of the 4th, John C. West, recalled "our line at times could hardly be called a line at all. . . It was impossible to make a united charge. The enemy were pretty thick and well concealed. It was more like Indian fighting than anything I experienced during the war."[30]

Private Roland Bowen and his comrades of the 15th Massachusetts were ordered forward from the main Union line with the 82nd New York to defend the Emmitsburg Road on July 2 and help protect the right flank of the 3rd Corps. As soon as both regiments reached the road, they went to work tearing down the fencing along and near the road to build a breastwork for protection. Because of a slight rise in the ground in front of their position, the men in the two regiments had a very limited view of the enemy in their front and had to rely on their skirmishers to warn them of an advance. The first notice that the enemy were approaching was when "all the pickets came rushing in, some kept straight for the rear, but we made most of them stop and form in with us." Bowen does not relate how they made them stop. Did they threaten to shoot the men? Did they physically restrain them or appeal to their courage? The line of Union soldiers made ready to receive the Confederate attack. "Some said they could see their heads [the Confederates] in the tall grass," wrote Bowen, "and the musketry fire commenced moderately." Bowen "looked but I could not see any thing of them, but I had plenty of ammunition, so I let fly into the grass." He also took out a handful of cartridges and laid them beside himself, as well as his ramrod "so as to grasp it in an instant." Both of these preparations were contrary to training, but Bowen noted "nearly all the boys done the same."[31]

Approaching the position of Bowen's regiment was Brigadier General Ambrose Wright's Georgia brigade, consisting of some 1,400 men. But instead of advancing walking upright in the 15th Massachusetts's front, Wright's men crawled through the tall meadow grass on their hands and knees. Bowen saw the grass moving and bayonets rising above it before he saw any live Confederate soldiers. When Wright's men had reached what they thought was a point close enough to rush the enemy, they fired a "deadly volley" at the Union troops, then "sprang forward with that demoniac yell." Bowen and his comrades replied with their own shout, rising up from their prone position to their knees and resting their rifles on their fence rail breastwork. "We give them one of the most destructive volleys I ever witnessed," wrote Bowen. The volley staggered Wright's men and Bowen thought they acted like they "lost their confidence," and "staggered and wavered slightly." But he detected no panic in the Confederate ranks. Both sides poured fire into one another as fast as they could load their weapons.[32]

The men of the 82nd New York probably had a similar experience to that of the 15th Massachusetts, except that they had no one protecting their immediate left flank and Wright's line extended beyond the front these two Union regiments could cover. Wright's men quickly took advantage of this situation and rushed to outflank the New Yorkers. Now exposed to both a front and flank fire, the 82nd fell back. This, in turn, allowed Wright's men to pour a flanking fire into the 15th Massachusetts. Bowen thought that the retreat of the 82nd inspired the Georgians in his front "with new courage," and again they advanced, loading and firing as they did so. "We poured one continual storm of lead on them, but they heeded not," wrote Bowen. By this point, from Bowen's perspective, "everything seemed to be in an utter state of confusion." His comrades, sensing that they would be unable to halt the Confederate attack, began to break for the rear, although Bowen and others were hollering and shouting such things as "nothing but cowards run." But it was to no avail and soon "the stampede became general." Bowen became a prisoner, along with twenty-five comrades. Many others were shot down during the retreat, which left them exposed in the open to Wright's pursuing men. The 82nd New York reported 153 killed and wounded out of some 350 men carried into the fight. Bowen's regiment did not report their losses for this day, but an estimate would place them at nearly 100. The entire action probably did not last more than ten to fifteen minutes.[33]

Bowen's descriptive and honest account of his experience on July 2 reveals some aspects of the nature of line-of-battle fighting. Both sides used cover as much as possible, contrary to the popular notion that Civil War soldiers generally stood in the open and blazed away at one another. The Confederates apparently made effective use of the tall meadow grass to get as close to the Union line as possible before exposing themselves. However, once they did expose themselves they were unable to overcome the Federal soldiers, who had the advantage of cover and sufficient numbers to produce enough firepower to hold Wright's men at bay. It was the threat to the flank that unhinged the Union line. Most of the Union casualties probably occurred when the two regiments retreated back to Cemetery Ridge. The men had to cross open ground exposed to the fire of much of Wright's Brigade. Conversely, most of Wright's casualties in this action probably were suffered in the first contact.

This engagement also highlighted the importance of training and discipline in the type of bloody fighting that characterized the line of battle. Despite the telling volley the 15th Massachusetts delivered into Wright's line, and the confusion it caused, Bowen noted that there was "no panic" among the Confederates. The bonds of discipline and leadership are likely what held the Confederates together and enabled them to endure their losses and drive their attack home. According to Bowen, Wright's soldiers "heeded not" their casualties in the final action that overcame the Union line along the Emmitsburg Road but pressed ahead in spite of them. In contrast Bowen noted that he did not see a single officer of his regiment once the fighting became general, which might have influenced the confusion Bowen described, and the resolve of the enlisted men. The action also illustrates that even though soldiers might think they had delivered a "murderous fire," most bullets missed their mark, even at close range. Though Bowen's regiment caused numerous casualties to the Confederate regiment opposite them, they did not inflict enough to demoralize the Confederates or discourage them from continuing to fight.[34]

Soldiers in the line of battle did not always take cover when engaging an attacking line, even when it was available. Captain Donaldson of the 118th Pennsylvania, whose regiment occupied a finger of woods bordering the Wheatfield on July 2, described his regiment's position "in all respects a good one ... with rocks and huge boulders scattered about forming ample protection." Yet when the line Donaldson's regiment and brigade occupied was attacked by Confederates of George T. Anderson's Georgia brigade, the captain observed that the men in his regiment were so eager "that I did not notice one of them taking advantage of the trees and rocks, but all standing bravely up to the work and doing good execution."[35]

The Confederates of Anderson's Brigade attacking Donaldson's regiment employed an effective tactic that apparently only the Confederates used at Gettysburg. This was to advance steadily loading and firing. Lieutenant J. C. Reed, who served in the 8th Georgia of Anderson's Brigade, wrote about this action that when his men advanced against the Union line they "came on almost at a run, firing vigorously." The effect of this fire was to drive many of the Union soldiers to cover. From Reed's perspective, they "were thinking more of shelter by the rocks and trees than of firing," which would have affected the accuracy of the Federal soldiers' return fire. Reed's regiment failed to drive the Federal line, mainly he thought, because of a bog that crossed their line of advance. To cross the bog, Reed wrote, meant "that we have to cease firing, and the men on the other side know their advantage."[36]

That this type of advance was not unique to the 8th Georgia is evident from an unofficial report of the battle by Colonel Joshua L. Chamberlain of the 20th Maine. In describing the advance of the 15th Alabama against his regiment, he wrote that the Alabamians made "what they call a 'charge'—that is, advancing & firing rapidly." What some Confederate commanders had apparently learned through experience is that once their line came within effective rifle-musket range of the enemy, permitting their men to advance loading and firing forced the enemy to keep their heads down and made their return fire less effective. It was a primitive form of marching fire that the U.S. Army would use very effectively in World War II.[37]

The experience of the soldier within the line of battle differed considerably from that of the skirmisher. While the skirmisher often experienced a feeling of isolation and separation from his comrades, the soldier in the line of battle could see many of his comrades around him, from whom he could draw courage and a feeling of security. At the same time, when a line began to dissolve, or it suffered many casualties, demoralization and panic could spread quickly through the ranks. An infantry line of battle was a noisy place, with the constant crash of hundreds of rifles being fired, officers shouting encouragement or orders to their men over the din, and file closers pushing and shoving their men around to fill in gaps in the line or to maintain a semblance of order. Lieutenant Charles A. Fuller of the 61st New York, who served as a file closer for his regiment during fighting in the Wheatfield on July 2, noted: "In battle the tendency is almost universal for the men to work out of a good line of battle into clumps. The men of natural daring will rather crowd to the front, and those cast in more timid or retiring molds will almost automatically edge back and slip in behind. Hence the necessity of not alone commissioned officers in the rear to keep the men out in two ranks, but sergeants as well."[38]

Maintaining order and communicating commands were constant challenges for the officers. During the fighting in the Wheatfield, Colonel John R. Brooke, whose Union brigade had halted in the middle of the grain field to engage Confederates of Brigadier General George T. Anderson's brigade, concluded that only a general advance of his brigade could succeed in dislodging the Confederates, who were under the cover of Rose Woods at the southern end of the field. Ordering the advance was easy. Executing it was not. One of Brooke's colonels, Daniel Bingham, commanding the 64th New York, recorded that "the men were firing as fast as they could load. The din was almost deafening. It was very difficult to have orders understood, and it required considerable effort to start the line forward into another charge." It can be imagined how difficult it would be without radio communication and while under fire to communicate even a simple command to enough officers to see it executed. In this particular situation, the two color bearers of the 64th ran several yards in front of the regimental line "so that they were dimly perceivable through the clouds of smoke." The regimental colors were among the most efficient means of communication in the confusion of combat, for the men were trained that where their colors went they should follow. This brave act by the 64th's color bearers caused the body of their regiment to start to advance, which in turn began a general advance of Brooke's entire brigade.[39]

The colors were an integral element of a Civil War line of battle. These flags served a functional purpose as a means of communication in battle, as described above, and as a mark to align the line upon. But they also served an inspirational purpose. The colors were the visible symbol of the regiment and as such they were a source of great pride among the soldiers of the regiment. Only the bravest men were entrusted to bear them in battle. When a unit was driven from the field, the colors inevitably served as the rallying point to reassemble the regiment. On the attack, a brave color guard could embolden the men of a regiment to continue forward or to hold their ground, despite the gravest danger and ghastly losses. The 26th North Carolina and 24th Michigan provided a notable example of this in their desperate battle on July 1. The 26th launched a frontal assault upon the 24th's position in Herbst Woods on the afternoon of that day. The combat that ensued was some of the most terrible of the entire battle. The two lines exchanged fire at a range of about 40 yards. In battle, officers and color bearers were the favorite targets of the opposing lines, partly because both could be conspicuous in the smoky atmosphere but also because it could be demoralizing for men to see their colors shot down. During the course of the fight between these two regiments, the 26th North Carolina had fourteen color bearers shot down, including their colonel, Henry K. Burgwyn, who was killed, and the 24th Michigan lost seven color bearers killed or wounded. For the North Carolinians, keeping the colors up in this battle became a symbol of their resolve to continue the assault despite appalling loss, and for the Michigan men, of their determination to hold together to defend their position.[40]

The historian of the 14th Connecticut recalled how Union small arms and canister fire during Pickett's Charge on July 3 caused the Confederate line in his front to waver and falter, "even their courage forbidding them to face such a storm of musketry." The Confederate color bearers and their color guard then advanced toward the Union line and "planted their battle flags in the ground much nearer," in order to encourage their regiments to continue the advance. They succeeded in encouraging many men of their regiments to move up to their position, but in this instance, courage could not overcome the firepower the Federals delivered from Cemetery Ridge, and the attack failed. In many instances along the line, the colors had been advanced so near, and at the Angle, into, the Union position, that the colors were captured when the attack failed.[41]

The fighting men in a line of battle quickly lost whatever neat and orderly appearance they might have started into action with. George K. Collins, a captain in the 149th New York, whose regiment was heavily engaged at Culp's Hill on both July 2 and July 3, recalled that the men's clothing by July 3 was "ragged and dirty," and "their faces black from smoke, sweat and burnt powder, their lips cracked and bleeding from salt-petre in the cartridges bitten by them, and while loading and firing for dear life, [they] resembled more the inhabitants of the bottomless pit then quiet peaceful citizens of the United States of America." The physical appearance was so changed that Collins noted, "The people at home would not have recognized their friends, and a father would have been perplexed to know his own son." He also added an interesting observation that on each man "every pocket was torn open and the contents lost in a manner which none could explain." Possibly the men had stuffed extra cartridges into their pockets and in their haste and excitement to reach for them had torn their pockets.[42]

In his candid reminiscence, Collins also described the reaction of the men to deaths of their comrades around them. "At first the killed were tenderly put back out of the way," he wrote, "but afterwards attention was given only to the wounded unable to get off the field without help. Occasionally the dead were tossed from under foot, but in most instances remained where they fell, and were sat upon by the men while loading their pieces." Captain Henry L. Abbott of the 20th Massachusetts also offered a perspective on this callousness of fighting men under fire when he wrote the father about the death of one of his dearest friends, Lieutenant Henry Ropes, who was killed by "friendly fire" during artillery action on the morning of July 3. Abbott stated that the men of Ropes's company "actually wept when they showed me his body, even under the tremendous cannonade, a time when most soldiers see their comrades dying around them with indifference." Clearly, Abbott saw this reaction of his veteran soldiers to be unique in battle. While actually under fire those not shot were simply too busy with trying to stay alive to expend emotional energy on the dead. That would come after the battle, when men could grieve a dead buddy or question why they were spared while others died. Abbott himself was a case in point. The death of his closest friend in the regiment, a man he thought of as a brother, did not affect his performance, for he behaved with great gallantry during the Confederate infantry assault that followed the bombardment. But after the heat of battle had subsided and Abbott had time for his own personal thoughts, he felt Ropes's loss deeply. "I can't cease to think of him, whenever I am alone," he wrote his father on July 6.[43]

Hand to Hand

Sensationalists delight in thrilling the public with tales and images of soldiers locked in hand-to-hand combat at Gettysburg, and it is a commonly held belief by the public that this type of fighting was commonplace during the battle. The film *Gettysburg* has helped reinforce this idea with its scenes of hand-to-hand fighting between the 20th Maine and 15th Alabama on Little Round Top, and a huge encounter at the climax of Pickett's Charge that almost resembles a giant rugby scrum. Hand-to-hand fighting did occur in both of these actions but not on the scale depicted, and these were the exception, not the rule in infantry combat at Gettysburg. Most infantry fighting at Gettysburg took place at ranges of 100 yards or less. The report of a Union 2nd Corps brigade commander, Colonel Norman J. Hall, is instructive. During the grand Confederate assault of July 3, Hall reported, "I caused the Seventh Michigan and Twentieth Massachusetts Volunteers to open fire at about 200 yards . . . The remainder of our line reserved its fire until within 100 yards, some regiments waiting even until but 50 paces intervened between them and the enemy." Although the Springfield or Enfield rifle, the small arm most soldiers in both armies carried, was accurate in the hands of a good rifleman at a range of 300 yards, and had a killing range nearly up to a mile, the low muzzle velocity of these rifles meant that at longer ranges they did not have a flat trajectory. A soldier firing at a target 300 yards away elevated his sights so that the bullet actually traveled in an arc to its target. If the firing soldier misjudged the range, it was possible for the bullet to pass completely over the enemy. Most soldiers were also not good shots, and the excitement of battle made their aim even more erratic. The greater the range then, the less the likelihood that a firing line would inflict enough damage on an opposing line to check its advance or cause its retreat. Consequently, most infantry officers preferred to allow an advancing enemy to approach to within 100 yards before opening fire. Veteran infantry officers also understood that the first volley or two from their line would be the most effective and well-aimed fire their men would deliver, for once the action became general the men invariably fired at will and the effectiveness and accuracy of their fired dropped off.[44]

Holding fire until the enemy were within short range increased the odds that an advancing line could be stopped, or a defending line dislodged, but it did not guarantee it. During Pickett's Charge, the 7th Michigan, one of Colonel Hall's regiments, opened fire at "short range" with what their Major Sylvanus W. Curtis reported as "terrible effect, mowing them down by scores." Yet despite this fire, Curtis related, "still they came on till within a few yards of us, when the order was given to fix bayonets." But it never came to a hand-to-hand encounter. The Confederate ranks were too thinned and disorganized to press the attack home and close with Curtis's regiment, and the Confederate soldiers retreated or surrendered. It was not unusual for an infantry action at Gettysburg to close to the distance related by Curtis—mere yards—and there not be any hand-to-hand fighting. Captain Henry L. Abbott of the 20th Massachusetts, after the repulse of Pickett's Charge, examined the ground held by the opposing sides and related that the "rows of dead after the battle I found to be within 15 to 20 feet apart, as near hand to hand fighting as I ever care to see." Abbott's statement is interesting on two points: first, it illustrates how close-quarter the fighting was, but it is also apparent that Abbott was surprised at how close the action was, and that he, a veteran of several large battles, had never seen or participated in hand-to-hand fighting. The famous clash between the Iron Brigade and Pettigrew's Brigade on July 1 in Herbst Woods, as related earlier, took place at distances of between 40 and 20 yards. That it never came to hand-to-hand had to do with the fact that Iron Brigade continually kept its ranks closed up despite extremely heavy casualties and maintained a steady fire along its front so that no one could cross that final 20 yards to close with them. Additionally, the Iron Brigade continually fell back to a new line when one of Pettigrew's regiments managed to gain a flanking fire.[45]

During the action on July 2, several regiments of Brigadier General

Joseph B. Kershaw's brigade advanced to the area of Stony Hill near the Wheatfield. Here, two brigades of Brigadier General John Caldwell's division from the 2nd Corps counterattacked them. Kershaw's men had good cover from rocks and trees and defended their position stubbornly. The Federals of Zook's and Kelly's brigades were equally determined on the attack. Lieutenant Colonel Franklin Gaillard of Kershaw's 2nd South Carolina, describing the action afterwards, wrote, "The enemy's infantry came up and we stood within thirty steps of each other. They loaded and fired deliberately. I never saw more stubbornness. It was so desperate I took two shots with my pistol at men scarcely thirty steps from me." Yet again, despite the close, almost point-blank range of this action, it never came to hand-to-hand, probably because the volume of fire Kershaw's men kept up made it impossible to get closer than 30 yards. Instead, the Federals maintained a heavy fire on Kershaw's front with part of their force and maneuvered the rest to get around the South Carolinians' flank. Kershaw kept swinging back the flank of his right regiment, the 7th South Carolina, until "the two wings of the regiment were nearly doubled on each other." There came a point in the fight, however, when Kershaw realized that if he attempted to hold his position any longer he might well be cut off and forced to surrender, or suffer devastating losses, and he ordered a retreat.[46]

Where hand-to-hand actions did occur at Gettysburg two factors were present. First, the fire of the defending unit was scattered or disorganized enough that the attacker was able to advance across the deadly ground— the final 20 to 40 yards—and close with the enemy, or the attacker was able to gain the flank or rear of the defender. Secondly, the defenders either believed they could fight off the number of attackers who closed to point-blank range, or they were left with no option and felt forced to engage in hand-to-hand combat. The most well known hand-to-hand fighting at Gettysburg occurred between the 20th Maine and 15th Alabama on Little Round Top. The film *Gettysburg* depicted this as almost general along the line of the two regiments as the 15th drove home its attack. In reality, the hand-to-hand combats in this battle were few, isolated, and of very brief duration. The fire of the 20th Maine did not break the initial charge of the 15th Alabama until the Confederates were "within a dozen yards" of the 20th's line. That the Confederates were able to get this close is probably explained by the fact that the 20th had extended into a single line to cover a broader front, thereby significantly reducing its firepower; there was cover that partially sheltered the Confederates in their advance; and the 15th Alabama was a well-disciplined, well-led unit. Having reached a position so close to the 20th, the Alabamians were able to muster rushes by what Colonel Chamberlain described as "squads of men," who "broke through our line in several places, and the fight was literally hand to hand."[47]

That the Confederates were able to do this had to do with the short distance they had to cover to reach the Union position, and again, because the 20th's single line had gaps in it due to casualties. To recover ground lost, or drive these Confederate thrusts back, squads of the 20th would launch counterattacks. Colonel William Oates, who commanded the 15th Alabama, recalled an encounter during one of these. "I, with my regiment, made a rush forward from the ledge. About forty steps up the slope there is a large boulder about midway the Spur. The Maine regiment charged my line, coming right up in a hand-to-hand encounter. My regimental colors were just a step or two to the right of that boulder, and I was within ten feet. A Maine man reached to grasp the staff of the colors when Ensign Archibald stepped back and Sergeant Pat O'Conner stove his bayonet through the head of the Yankee, who fell dead. I witnessed that incident, which impressed me beyond the point of being forgotten." It is unlikely that Oates led his entire regiment up the slope, as he states, or that the entire 20th Maine charged his line, simply because it was impossible at that stage of the action, due to the smoke, terrain, confusion, and enemy fire, to get an entire regiment to act in unison. Oates's thrust more likely consisted of a company or so, just as the Union counterthrust that led to the hand-to-hand action was not a general advance by the 20th, but rather a local attack to drive the Confederates back. Of all the incidents Oates probably observed during this action, he chose to relate the story of Sergeant O'Conner, indicating that it both shocked the colonel and was a rare, isolated event.[48]

The 20th Maine's famous bayonet charge at the critical point in the action between these two regiments resulted in relatively few personal encounters. Chamberlain had one in which a Confederate officer attempted to shoot him at point-blank range, but missed. Chamberlain placed his sword at the officer's throat and the fellow quickly surrendered. Most of Oates's men withdrew in the face of the 20th's bayonet charge. The rest, who were either too exhausted to flee or wounded, surrendered.[49]

While Henry Abbott related that the fighting near the "clump of trees" at the climax of Pickett's Charge was "as near hand to hand fighting as I ever wish to see," there was a brief hand-to-hand encounter during this action that Abbott did not witness. When the Confederates of Garnett's and Armistead's brigades swept up and forced several companies of the 71st Pennsylvania back from the Angle, it exposed the right flank of the 69th Pennsylvania, who defended the stone wall on the left of these 71st companies. In an effort to exploit this small advantage gained, Brigadier General Lewis Armistead led a contingent of men over the wall past the flank of the 69th in an effort to gain the flank of the Union troops defending the recessed wall, east of the Angle, or to penetrate to the rear of Gibbon's division, who held the line from the Angle south. To protect their flank, the right three companies of the 69th were ordered to change front to the right, which meant get up from their position at the wall and form a new line at right angles to their old one. The two far-right companies managed to execute this movement, but the company commander of the third company in line, Company F, was killed and his company did not move. As a result a gap developed between Companies F and A, which had already changed front and was firing at Armistead and his group. There were still many Confederates at the wall near the Angle who had not joined Armistead, and a strong group of them now rose and rushed into this gap. They fell upon Company F from its flank and rear so quickly that every man was either killed or captured.

Company D was on F's left flank and their commander, Captain Patrick Tinen, pulled his men out of the line to confront the Rebels who had overrun their comrades and threatened to unhinge the entire regimental line. For a few brief moments men closed and fought hand-to-hand. Anthony McDermott, a member of Company I, recalled that Private Hugh Bradley, "who was quite a savage sort of fellow, wielded his piece, striking right and left, and was killed in the melee by having his skull crushed by a musket in the hands of a Rebel." Another private in Company D, Thomas Donnelly, "used his piece as a club" and struck to the ground a Confederate who attempted to take him prisoner. But lest we think that everyone in Company D was wielding his musket like a club, McDermott noted in a letter about this part of the battle, "I cite these two instances because they seem to have been particularly noticed by their comrades and spoken of by many of the men of the above mentioned Co." In other words, clubbed muskets were the exception, not the rule, in this close-quarter fight. No one used bayonets, for no one in this area had them fixed. The greatest killer in this fight was the rifle used at point blank range.[50]

The only bayonet fighting of real note during the entire battle occurred during the collapse of Union resistance in the Wheatfield. While standing in the Wheatfield during a meeting of Union and former Confederate officers in 1869, Confederate Brigadier General William T. Wofford made the chilling statement that "more men were killed here with the bayonet than he had ever known before in the war." Wofford was not referring to the fighting in the Wheatfield in general, but to one specific fight that his brigade engaged in with Colonel Jacob Sweitzer's brigade of Barnes's 1st Division, 5th Corps. During the late afternoon fighting in the Wheatfield on July 2, Sweitzer's brigade advanced to the southern end of the Wheatfield to support Colonel John R. Brooke's brigade of Caldwell's 1st Division, 2nd Corps. Sweitzer faced south, because that was where Brooke was. Two other brigades of Caldwell's division, Zook's and Kelly's, supported his right flank, holding a place called Stony Hill. But the advance of Wofford's Brigade, along with elements of Kershaw's Brigade, drove Zook and Kelly off the hill and threatened the flank and rear of Sweitzer. To meet the emergency Sweitzer attempted to change front from the south to the west with the 62nd Pennsylvania and 4th Michigan. As these regiments tried to carry out this maneuver, Wofford's men emerged from the woods bordering the Wheatfield and engaged the Federals at close-quarters. Jacob A. Funk was a color bearer in the 62nd Pennsylvania, and he recalled, "The battle now raged in all its fury as foe grappled with foe and the Bayonet was freely used." Soldiers also wielded their rifles as clubs. Funk recalled a Confederate officer who demanded that the Pennsylvanians surrender, "when one of my company turned round and clubbing his musket brought the but down on the officers hed smashing him down on the spot."[51]

In the 4th Michigan, on the 62nd's right, a hand-to-hand conflict swirled around the colors of the 4th. The color bearer of the regiment was shot and a Confederate soldier picked up the fallen flag and was bearing it away when Colonel Harrison Jeffords, the 4th's commander, rushed after the exultant enemy and ran him through with his sword, killing him. A Confederate behind Jeffords bayoneted the colonel, inflicting a mortal wound, whereupon a 4th Michigan lieutenant shot Jeffords's assailant with his pistol. While some men might have been killed by the bayonet in the 62nd Pennsylvania or in Wofford's Brigade, it is interesting to note that a lieutenant in the 4th recorded that Jeffords was the only man in his regiment killed by a bayonet thrust. This would indicate that while Wofford's statement that he observed more men killed with the bayonet than he had ever known before might have been true, the number actually killed was not very high, even though it is clear that this action was a genuine hand-to-hand fight. But, like the fight of the 69th Pennsylvania described above, bullets cut down many more men in this action than did bayonets or clubbed muskets.[52]

Artillery

Unlike the infantry, which could fight in literally any terrain, artillery used direct fire weapons that needed a clear line of sight to their targets. Artillerists also needed open ground to unlimber their field pieces, and room to place the limber and caisson that accompanied each gun. Artillery at Gettysburg fell into two categories, rifled guns and smoothbores. The predominant rifled pieces were the 3-inch Ordnance Rifle and the 10-pound Parrott Rifle. Each had an extreme range of about 2 miles, although they were more effective at ranges of 1¾ mile and below. The 12-pound Napoleon was the most common and favorite smoothbore field piece in both armies. It had an extreme range of about one mile. Both rifled and smoothbore artillery fired four types of ammunition: shot, shell, case or shrapnel, and canister. Solid shot consisted of a solid ball or bolt that was good for demolishing buildings or breastworks, dismounting guns, and creating mayhem in large infantry formations. Shell consisted of a hollow shell with an explosive charge detonated by a timed fuse that burst, one hoped, over or in the midst of its target, shattering the shell into several large iron fragments. Shrapnel or case shot was the most effective long-range ammunition in the artillery's arsenal. Like shell, it was a hollow shell with an explosive charge within the shell casing, but it contained a number of small iron shrapnel balls that scattered when the shell burst. With their greater range and accuracy, the rifled guns were more effective in delivering this ammunition to the target than smoothbores. But the smoothbores reigned supreme when firing canister. A large tin can filled with iron balls, canister was like a big shotgun shell. Canister for the smoothbores carried more canister balls than did canister for rifled artillery, and its pattern was more cylindrical, and more deadly. Its maximum range was approximately 600 yards, which meant that when enemy infantry were in canister range, the gunners were themselves in range of the infantrymen's rifles.

Although artillery was not the great killer on the battlefield that it would become in the late-nineteenth and twentieth centuries, it still could inflict considerable punishment on an infantry line of battle, and sow disorder in its ranks, often taking the steam out of an attack or demoralizing a defensive line. But for artillery to be effective against enemy infantry, it required friendly infantry support. If a battery were left unsupported, the opposing infantry could spread out in skirmish formation to attack it. With friendly infantry in support, enemy infantry were obliged to remain massed to concentrate their firepower. This gave artillery units a target they could damage.

Artillery served two purposes at Gettysburg and other battlefields of the war. First, its mission was to damage, silence, or drive off opposing artillery. Second, and of equal importance, was to damage enemy infantry by killing and maiming, and by sapping morale. The two major infantry attacks at Gettysburg, by Longstreet's Corps on July 2, and the Pickett-Pettigrew-Trimble assault of July 3, were both preceded by heavy massed artillery bombardments. Some 62 Confederate guns participated in the bombardment on the 2nd, while on the 3rd nearly 150 took part. Their purpose was to pave the way for the assaulting infantry by silencing the Federal batteries, smashing up the defensive formations of the defending infantry, and creating as much confusion as possible. In both instances, Union artillery responded to the Southern guns, with the goals of silencing them if possible and of drawing fire away from the infantry.

Artillery duels often were not bloody affairs, for when a battery got the range of its opponent, the opponent would generally limber their guns and move to a new position. But the two big gun duels, on July 2 and July 3, as well as one that took place on the afternoon of July 2 between Union batteries on East Cemetery Hill and Major J. W. Latimer's battalion of Edward Johnson's Division on Benner's Hill, proved very costly in men and horses for some batteries. Colonel Edward P. Alexander, who served as Longstreet's de facto chief of artillery on July 2 and July 3, recalled that he expected the bombardment he arranged on July 2 to be brief and decisive because the range was so close (about one-half mile). "But they [the Federals] really surprised me, both with the number of guns they developed, & the way they stuck to them," he wrote. Alexander thought the artillery duel on the 2nd the hardest and sharpest artillery fight of the war.

One of his batteries lost 36 killed and wounded out of 71 men, a 50 percent casualty rate, which was staggering for an artillery battery. His own battalion lost 139 men and 116 horses on July 2nd and 3rd, with two-thirds of the losses coming on the 2nd. These unusually high casualties were all inflicted by enemy artillery fire. Losses over 20 or 30 percent could cripple an artillery unit, since most casualties occurred in the gun crews. Some idea of the severity of this artillery duel can be gained when it is understood that in a hard-fought duel losses were more typically around 5 or 6 percent.

The experiences of two Union batteries during the artillery bombardment and duel of July 3 offer some idea of the nature of the combat when artillery faced opposing artillery. These were Battery A, 4th United States Artillery, commanded by Lieutenant Alonzo H. Cushing, and Battery B, 1st Rhode Island Light Artillery, commanded by Lieutenant Walter S. Perrin. These two batteries were positioned immediately north and south of the celebrated "clump of trees" on Cemetery Ridge, near the point where the Confederates hoped to make a lodgment in the Union line with their massed infantry attack. Therefore both of these batteries, as well as all the other batteries in the 2nd Corps on Cemetery Ridge, were the especial targets of the Confederate preattack bombardment. When the Confederate batteries opened fire at about 1 P.M. that afternoon, the men of both batteries immediately moved to their posts, officers and gunners taking position near their guns, and drivers mounting their horses (three per gun and limber) on the limbers and caissons. Within minutes the din of the enemy artillery fire "was one grand raging clashing of sound," with the "bursting of shells so incessant that the ear could not distinguish the individual explosions." A nearby infantryman described the shelling as if "all the Demons in Hell were let loose, and were Howling through the Air." There were nearly 150 Confederate artillery pieces shelling the Union line. If they were firing one round a minute, which in the early stages of the bombardment they may have been, then theoretically *every second* at least two shells were striking or exploding on the Union lines.[53]

In Battery B, Albert Straight, a sergeant in the battery, wrote afterwards that the Confederate fire "was terrible beyond description; the air was full of shell hissing and bursting. They came so fast there was no dodging them." Three shells struck the gun he commanded before they burst. One of these burst at the muzzle of the gun, and its effect illustrates the shocking wounds artillery fragments could inflict. The no. 1 gunner,

William Jones, had finished swabbing the barrel and took his place between the gun barrel and wheel on the right side. Alfred G. Gardner, the
no. 2 on the gun, took his place between the muzzle and left side wheel
and was in the act of inserting the ammunition when the Confederate
shell struck the muzzle and exploded. A fragment cut the top of the left
side of Jones's head completely off, killing him instantly. Gardner took
a fragment in the left shoulder, which nearly severed the arm. He died
shouting, "Glory to God! I am happy! Hallelujah!" One can only imagine
the potentially demoralizing effect such ghastly wounds could have upon
the men who witnessed them. This may have been what happened to the
no. 4 gunner on the piece. According to the corporal of the piece, J. M.
Dye, after the death of Gardner and Jones the no. 4 "was played out and
lay on the ground." Dye tried to get him up "but he wouldn't budge." Was
it exhaustion that prostrated no. 4, or terror?[54]

In Cushing's Battery A, a gunner of that command, Christopher Smith,
recalled, "men and horses were being torn to pieces on all sides. Every few
seconds a shot or shell would strike right in among our guns," recalled
Christopher Smith. An infantryman nearby wrote that the bursting of
shells was "so incessant that the ear could not distinguish the individual
explosions." Besides shot, shell fragments, and shrapnel balls that came
hissing and screaming through the air, fragments of rock from the stone
wall in front of Battery A, 1st Rhode Island (on Cushing's right) were
knocked loose by the impact of the Confederate ordnance and came "flying through the air" like shell fragments. Wounded horses screamed and
plunged in their traces. Dead ones dropped down and had to be cut out
by the limber drivers. Men who were struck by shell fragments or solid
shot were frightfully mangled. Arsenal H. Griffin served with the battery's limbers. One enemy shell struck two horses on one of the limbers.
It passed clean through the first horse and exploded within the second.
A fragment ripped into Griffin's abdomen. Christopher Smith saw him
writhing in pain on the ground. His intestines were spilling out of the
wound and Griffin begged his comrades to shoot him. When no one did,
Griffin pulled his revolver and shot himself in the head to end his misery.
The scene that unfolded around Cushing was not simply frightening, it
was terrifying.[55]

In an example of the contagion of fear, about fifteen minutes into the
bombardment a solid shot struck the no. 3 gun and tore away a wheel. Sergeant Thomas Whitston, who commanded the gun, panicked. Whitston

was a good soldier, but everyone has a breaking point, and perhaps this close brush with death was Whitston's. He started to run and his crew followed. Cushing reacted swiftly, drawing his revolver he shouted at Whitston, "Sergeant Whitston, come back to your post." Then he added so everyone who could hear him over the din understood his intent, "The first man who leaves his post again I'll blow his brains out." Would Cushing have shot Sergeant Whitston? Perhaps, but we shall never know, for the threat of force and sharp words of command stopped the sergeant in his tracks and helped him regain his composure. He led his crew back to the caissons, retrieved the spare wheel and soon had his gun firing again.[56]

Fifteen minutes after Cushing's, Perrin's, and the other Union batteries began to respond to the Confederate guns, the dense smoke produced by each discharge completely obscured observation of targets. Gun crews estimated the range and elevation and blazed away. So, the longer the bombardment continued the less effective it became. E. P. Alexander reported that he was only able to judge the effectiveness of his batteries' fire by the return fire of the enemy, as it was impossible to see what damage was being done.[57]

The bombardment's smoke screen convinced George Meade and Union Chief of Artillery, Henry Hunt, that they were simply wasting ammunition by continuing with counter-battery fire and they independently ordered the artillery to cease fire to conserve ammunition for the expected infantry assault. Cushing's and Perrin's batteries were engaged for nearly one and one-half hours when the artillery exchange finally died away. Perrin's Battery B was effectively destroyed in the bombardment and withdrawn. Cushing lost sixty-five horses killed and forty-one enlisted men and officers killed and wounded, most of them from his gun crews. By the end of the bombardment he could crew only two guns. Such severe losses from enemy artillery fire were not rare at Gettysburg, but they were uncommon during the war, and it took a particularly heavy concentration of artillery fire to achieve such destruction.[58]

Artillery generally proved far more effective than both line of battle or skirmishing and sharpshooting in producing death and destruction when trained on enemy infantry, except in instances where the infantry were under cover. This was the case in the bombardment before Pickett's Charge. Despite the vast quantity of ordnance the Confederates fired at the Union lines, losses among the Union infantry were relatively light. Part of the reason was the inferiority of Confederate ammunition. Many of their shells burst before reaching their target or were duds. The dense smoke that obscured vision also contributed to inaccurate fire. Civil War artillerymen had to see their targets to be able to hit them, and as one Union officer commented, "it is very difficult to hit a single line of troops," even without thick smoke. So much Confederate iron overshot Cemetery Ridge during this bombardment that Captain Henry Abbott of the 20th Massachusetts thought the Confederates were deliberately shelling the rear "with the intention of disabling the batteries & the reserves which they supposed to be massed in the rear of the batteries, in the depression of the hill."[59]

Although the July 3 bombardment inflicted few casualties, it did succeed in striking terror into the hearts of the infantryman on the receiving end. A sergeant in the 1st Minnesota recalled of the bombardment, "We had been badly scared many times before this but never quite so badly as then." This statement, from a man whose regiment suffered over 65 percent casualties in infantry fighting on July 2, but relatively few in the bombardment, offers some taste of the fear artillery fire could produce. Anthony McDermott, a soldier in the 69th Pennsylvania, observed that for two hours during this bombardment, "we did not enjoy any relief from the dread of being ploughed into shreds." McDermott struck upon the artillery's psychological impact. Artillery fragments produced ghastly wounds compared to small arms fire and this was unnerving. "It requires less nerve to face the enemy man-to-man in open field, than to lie down supinely while he hurls his missiles," wrote a 19th Maine captain, even though artillery fire produced fewer casualties than infantry fire.[60]

When infantry exposed itself in attacking, or defended without adequate cover, artillery could cause significant damage. As they were moving into an attack position on July 2, a shell found its mark in the 1st Texas infantry, killing and wounding fifteen men. Lieutenant George W. Finley of the 56th Virginia in Pickett's Division recalled that during their advance on July 3, as the Confederate units changed the direction of their attack some Union batteries were able to get an enfilading fire on their right flank, and "whenever it struck our ranks was fearfully destructive—one company a little to my right, numbering 35 or 40 men, was almost swept, 'to a man,' from the line by a single shell." During its attack on July 3, the left wing of Kershaw's South Carolina brigade received a mistaken order to move by the right flank at the very moment they were advancing upon a line of Union artillery. This meant that the regiments halted their advance, and each man turned to the right, presenting his side to the Union gunners. When they did so the Union batteries poured canister and shrapnel into their ranks. One officer wrote afterward: "The consequences were fatal. We were, in ten minutes time or less time, terribly butchered . . . I saw half a dozen [men] at a time knocked up and flung to the ground like trifles. It about that short space of time we had about half of our men killed or wounded." Captain Abbott, in the 20th Massachusetts, believed that had all the 2nd Corps batteries still been intact when the Confederate infantry appeared for the grand assault upon the Union center on July 3 "the rebels would never have got up to our musketry, for they were obliged to come out of the woods & advance from a half to ¾ of a mile over an open field & in plain sight."[61]

The key for infantry to avoid heavy loss from artillery was to move quickly so that the gunners had difficulty getting the proper range. During its charge on East Cemetery Hill on the early evening of July 2, Hays's and Avery's brigades of Early's Division encountered what one member described as a "perfect storm of grape, cannister [sic], schrapnel, etc." But Hays's Brigade in particular moved too swiftly for the Union gunners and passed through the most exposed point of their attack with relatively light losses. Hays believed that had the attack taken place in full daylight the result would have been a "horrible slaughter."[62]

Artillery that lacked infantry supports, or was inadequately supported, rarely could stop a determined infantry attack. On July 2, Confederate infantry overran over forty Union guns, although the Federals recaptured most of them before the day ended. The problem for a battery without infantry support was twofold. While a battery could protect its front with canister, its flanks were highly vulnerable. This is how the 21st Mississippi captured four guns of the 9th Massachusetts Battery on July 2 near the Trostle Farm. Lieutenant Richard S. Milton, an officer in the battery, reported afterward that although the center of the 21st Mississippi was "badly broken" by the battery's canister, "its flanks closing in on either side of us, obtained a cross-fire, which silenced the four pieces on the right, and prevented their withdrawal from loss of officers, men, and horses." Once enemy infantry were able to close to within effective small arms range a battery was at a decided disadvantage, for the gun crews and horses were greatly exposed. A battery had little choice if this happened but to attempt to limber and withdraw. Depending upon how close the enemy was this was a desperate effort, for during the process of limbering the battery could not defend itself. Most of the Union guns temporarily captured by the Confederates on the 2nd were taken while trying to limber to the rear. This was the case with Battery B, 1st Rhode Island Light Artillery on July 2. They were posted on a rock outcropping in advance of the main Union line supporting two regiments of infantry in the Emmitsburg Road when an attack by Ambrose Wright's Georgia brigade drove off the infantry and bore down upon the guns. The battery historian recalled that "by their exposed position the battery received the concentrated fire of the enemy, which was advancing so rapidly that our fuses were cut at three, two, and one second, and then canister at point blank range, and, finally, double charges of canister were used." Yet this fire failed to check Wright's Georgians and the battery was compelled to limber up and attempt to escape. Four guns managed to get off, but the Confederates shot the horses on the last two limbers and temporarily captured the two guns they were pulling.[63]

Cavalry

Nearly 12,000 Union and 6,600 Confederate cavalry participated in the Battle of Gettysburg. During the campaign, the Army of the Potomac had over 15,000 cavalry, while the Army of Northern Virginia fielded over 12,000 officers and men. By the time the two armies locked in combat at Gettysburg, part of both cavalry forces were detached from the main army, which accounts for the difference in the figures between the battle and the campaign. Of the 18,600 cavalrymen who participated in the battle, 610 Union and 286 Confederate became casualties, a casualty rate of 4 percent. This amounted to 2 percent of the total Union loss in the battle, and no more than 1 percent of the Confederate loss. Clearly, cavalry was not the decisive arm on a Civil War battlefield as it had been in earlier nineteenth-century wars. But these numbers are somewhat deceptive. During the entire Gettysburg Campaign the infantry and artillery of the two armies were in direct contact with each other for only a few days, while the cavalry were in contact with enemy cavalry on an almost daily basis. The result was that cavalry units rarely suffered large numbers of casualties (as compared to the infantry) in a single combat, but their cumulative losses of many days of contact with the enemy could be substantial. Brigadier General John Buford's division of the Army of the Potomac Cavalry Corps, for instance, reported only 176 casualties in the two brigades engaged on the main battlefield at Gettysburg, but if the engagement of Buford's 3rd Brigade at Fairfield, Pennsylvania on July 3 is added in, as well as all the other skirmishes and engagements the division took part in up to July 21, Buford suffered 714 casualties, about 14 percent of his reported strength as of June 30.[64]

Cavalry's principle roles on a battlefield such as Gettysburg were harassment, reconnaissance, and flank security, all of which their mobility and weapons ideally suited them for. They were the light forces of both armies, whose most important work was often performed before a major engagement and in its aftermath, when one army attempted to retreat.

There were three distinct types of cavalry encounters at Gettysburg: dismounted cavalry vs. infantry, mounted cavalry vs. infantry, and cavalry vs. cavalry. Because of the range and accuracy of the rifled musket, mounted cavalry rarely engaged enemy infantry. The one instance of this at Gettysburg, which will be discussed later, demonstrated the danger and futility of employing these tactics. Union Brigadier General John Buford, who commanded the 1st Division of the Army of the Potomac Cavalry Corps, demonstrated the most effective and cheapest (in terms of human lives) method of fighting infantry at the opening of the battle on July 1. When Buford was confronted by the advance of Heth's Division on the Chambersburg Pike on the morning of the 1st, he dismounted most of one of his two brigades. These were not novel or new tactics. They were dragoon tactics that the U.S. Army had used for years on the western plains. The troopers operated in teams of four. When they dismounted, every fourth man took his horse and the horses of his three comrades to a sheltered point in rear of the firing line. The other three troopers deployed in skirmish order with their carbines—in Buford's case, a single-shot breech-loading weapon capable of a higher rate of fire than the muzzle-loading infantry rifle, but with about one third the effective range. This latter point is important, for it meant that although dismounted cavalry could deliver a high rate of fire they were outranged by the infantry rifle, and the rate of fire was not so much greater that a cavalry skirmish line could hope to make a determined stand against an infantry line of battle. Buford's intent was not to stop Heth, but to harass him and force him to deploy from marching column to line of battle. This cost Heth time, which was Buford's objective. But Buford's delaying action has become the stuff of legends at Gettysburg, with some writers asserting that his troopers held off Heth for nearly two hours in bloody fighting. Buford did delay Heth for nearly two hours that morning, but not by slugging it out with him. He did it by using the mobility of cavalry to harass Heth, who did not have any cavalry himself, causing him to move cautiously. Colonel Birkett Fry, who commanded the 13th Alabama Infantry of Archer's Brigade, which directly confronted Buford's troopers, responded in 1878 to the question of how stoutly the cavalry resisted his brigade's advance on July 1. "As to the resistance made by the cavalry to the advance of our brigade on the morning of the 1st, I remember that it was inconsiderable, and did not delay us. Only small parties appeared in our front, and though I observed some instances of individual gallantry[,] they did us no damage."[65]

Other members of Archer's Brigade offered similar testimony. E. T. Boland, also of Fry's regiment, recalled that "just before Willoughby Run, the cavalry began to get stubborn, and our line passed the skirmish line. Then we drove them back until we crossed the Run and went up a short hill." The major of the 5th Alabama Battalion, which constituted most of Archer's skirmish line that morning, wrote immediately after the battle: "On the 1st my Battalion was deployed as skirmishers and lost only 7 men wounded although we drove the cavalry pickets and skirmishers of the enemy over three miles." Clearly, a fight that rambled over three miles and only resulted in seven casualties was, as Fry observed, "inconsiderable," particularly in light of the fact that his brigade shortly became involved in a fight with Union infantry, where their casualties numbered in the hundreds.[66]

Later in the afternoon of July 1, Buford dismounted parts of three regiments to support a stand by Union infantry of the 1st Corps on Seminary Ridge. This time the range was shorter and the cavalry line more concentrated and they exacted a terrible toll from the two South Carolina regiments that attacked them. The South Carolinians drove the dismounted Federals from their position, but it cost them some 61 dead and 210 wounded. Buford's entire division suffered only 127 casualties on July 1. This engagement demonstrated that dismounted cavalry with breech-loading carbines could do significant damage against infantry, but this was the only encounter at Gettysburg in which they did so. The other times that dismounted cavalry fought infantry, the fighting resembled Buford's morning skirmish, a harassing action with light casualties on both sides.[67]

There were few instances where mounted cavalry engaged infantry during the battle. The most famous was the charge of Brigadier General Elon Farnsworth's brigade in the rugged terrain near Big Round Top on July 3. Farnsworth's commander, Judson Kilpatrick, was an aggressive soldier who arrived on the far left flank of the Army of Potomac that day with orders to strike the enemy right and rear. He ordered Farnsworth to do just that despite a sketchy knowledge of enemy strength and dispositions, and terrain highly unsuited to mounted cavalry operations. Farnsworth moved forward with three regiments. Initially, only what amounted to a Confederate skirmish line with some supporting artillery confronted them, but the infantry were well concealed behind stone walls, brush, and timber. A member of the 1st Texas Infantry described Farnsworth's advance upon his position: "They rode down our skirmishers & charged us, and in a few seconds were on us, our Boys arose and pitched into them. They went through us cutting right & left, the firing for a few minutes was front, rear & towards their flanks. In a few minutes, great numbers of riderless horses were galloping around, & others with riders on were trying to surrender."[68]

Farnsworth's regiments charged in column of fours, rather than line, probably due to terrain, but such tactics gave cavalry considerable shock power on a narrow front. One of Farnsworth's regiments, the 18th Pennsylvania Cavalry, charged up nearly upon the line of the 1st Texas when, a captain of the 18th related, "all of a sudden the Rebs in our front appeared by the thousands [the 1st Texas actually had some 196 men spread out with some five or six steps between each man]. They seemed to come out of the ground like bees and they gave us such a rattling fire we all gave way and retreated towards the woods." The 18th lost five: two killed, two wounded, and one prisoner in the action. The whole affair was over in a matter of minutes.[69]

The two battalions of the 1st Vermont Cavalry who participated in the attack penetrated a soft point in the Confederate skirmish line but the vulnerability of cavalry to infantry with rifles, where the infantry had good cover and were not demoralized, made itself felt as the Confederates shifted two Alabama regiments to contain the breakthrough. That the infantry sensed their advantage over cavalry in this type of encounter—where terrain favored them—was reflected in the excited shout of a 4th Alabama lieutenant who sang out to his comrades as they closed on the Vermonters, "Cavalry, boys, cavalry; this is no fight, only a frolic; give it to 'em." Over fifty troopers were killed, wounded, or captured, including Farnsworth (who was killed). The charge accomplished nothing and it is tempting to point to this as further proof that mounted cavalry charges against infantry were pure folly by this stage of the war. But Kilpatrick sent Farnsworth into a situation ripe for disaster: rough, broken terrain intersected with stone walls, defended by infantry who may have been weak but who were not demoralized. In the fall of 1864 a powerful Federal cavalry would prove highly effective in mounted attacks upon Confederate infantry in action at Winchester and Cedar Creek. The difference was that in these actions the cavalry were committed after the Confederate infantry were disorganized by earlier fighting, and over terrain that allowed the cavalry to take advantage of its speed and mobility. The rifle had not made cavalry charges completely obsolete, but it had reduced the number of opportunities where a mounted charge offered hope of success.[70]

There was only one cavalry vs. cavalry encounter during the course of the battle on what is considered the main battlefield area; this was the large cavalry battle at what is known as the East Cavalry Field, but an argument can be made that the cavalry fights at Hunterstown on July 2 and Fairfield on July 3 were also a part of the main battle. However, we shall consider only the East Cavalry Field battle to form an idea of the nature of the cavalry vs. cavalry encounter. The battle pitted elements of four Confederate brigades against three Union brigades. Contrary to what might be imagined, much of this action consisted of skirmishing by dismounted troopers and shelling by horse artillery, punctuated by occasional mounted charges and countercharges. The accounts of two participants, one Union, one Confederate, shed some light on the nature of the fight. Lieutenant George W. Beale of the 9th Virginia Cavalry penned the following description of his regiment's part in the battle.

> The mounted men of our brigade were now ordered to charge. They passed through the yard of the barn [Rummel barn], under a raking fire from the guns to our right, and, doubling the head of the bottom, dashed up the slope to meet the foe. The little band led by Chambliss did not apparently exceed two hundred men. Reaching a fence which separated them from the enemy, they halted in line, and used their carbines until the fence was thrown down. It seemed to one who stood in a place of comparative safety that the enemy

slackened their fire curious to see if so few would dare to cross sabers with them. When the fence had been thrown down the brigade, with headlong impetuosity, hurled its columns upon the enemy's line, and for a few moments sabres flashed and pistols cracked. The work was soon over. Pierced and doubled up from center to flanks, the enemy fled in disorder, leaving many prisoners in the hands of our men. Meanwhile fresh troops of the enemy were dashing to the rescue, and our brigade, threatened in the rear, had in turn to fly. The captors of prisoners became now prisoners themselves.[71]

Captain James H. Kidd of the 6th Michigan Cavalry in George A. Custer's brigade offered the perspective of a Union horseman.

Just then a column of mounted men was seen advancing from our right and rear, squadron succeeding squadron, until an entire regiment came into view, with sabers gleaming and colors gaily fluttering in the breeze. It was the Seventh Michigan, commanded by Colonel Mann. Gregg, seeing the necessity for prompt action, had ordered it to charge. As it moved forward and cleared the battery, Custer drew his saber, placing himself in front, and shouted, "Come on, you Wolverines!" The Seventh dashed into an open field and rode straight at the dismounted line, which, staggered by the appearance of this new foe, broke to the rear and ran for its reserves. Custer led the charge half way across the plain, then turned to the left; but the gallant regiment swept on under its own leaders, riding down and capturing many prisoners.

There was no check to the charge. The squadrons kept on in good form. Every man yelled at the top of his voice until the regiment had gone, probably 1,000 yards straight toward the Confederate batteries, when, by some error of the guide of the leading squadron, the head of the column deflected to the left, making a quarter turn, and the regiment was hurled headlong against a post and rail fence that ran obliquely in front of the Rummel barn. This proved for the moment an impassable barrier. The squadrons coming up successively at a charge, rushed pell mell upon each other and were thrown into a state of indescribable confusion; though the rear companies, without order or orders, formed left and right front into line along the fence and pluckily began firing across it into the faces of the Confederates, who, when they saw the impetuous onset of the Seventh thus abruptly checked, rallied and began to collect in swarms upon the opposite side. Some of the officers leaped from their saddles and called upon the men to assist in making an opening ... The task was a difficult and hazardous one, the posts and rails being so firmly united that it could be accomplished only by lifting the posts, which were deeply set, and removing several lengths at once. This was finally done, however, though the regiment was exposed, not only to a fire from the force in front, but to a flanking fire from a strong skirmish line along a fence to the right.[72]

Kidd's regiment made their way through the gap in the fencing, with the Confederate troopers falling back before their advance. The charge continued until another fence was encountered and Kidd and his comrades observed a Confederate regiment, which had thrown down sections of yet another fence, forming to charge them. Without waiting to receive this attack, the 7th fell back.

The mounted action in both Beale's and Kidd's cases resulted because the enemy committed mounted forces to attack dismounted troops. Their regiments in turn were sent forward to meet these advances. In both cases they drove the enemy but became disordered, were countercharged by fresh enemy cavalry, and were forced to retreat. Beale's 9th Virginia did become involved in a brief hand-to-hand fight with sabers and pistols, but he states that it lasted only "a few moments." Kidd's regiment apparently recognized that in their disordered state, engaging a well-formed enemy courted defeat and they withdrew before sabers were crossed. Most of the mounted fighting these men described consisted of troopers riding back and forth and firing pistols and carbines at one another across a fence that separated them.

Beale's account and those of other participants of the battle reveal that mounted forces were highly vulnerable to flank attacks or attacks upon their rear. A mounted unit simply could not change its front to meet such an attack if it came suddenly. This is precisely what happened to the final Confederate charge of the action, when regiments of Wade Hampton's and Fitz Lee's brigades attempted to carry the day with a grand mounted charge. Although outnumbered at the point of contact, the mounted Federal cavalry that met this charge turned it back by aggressively striking Hampton's and Lee's front and flanks. This fight too, much of it fought with sabers and pistols, was brief, lasting only minutes, before the Confederate horsemen began to fall back. For all its apparent ferocity (at least as some participants described it), the clash resulted in light casualties for the numbers engaged. Total casualties in the entire cavalry battle were 181 Confederate and 254 Federals, or about 2 percent of the Confederate and 5 percent of the Union forces on hand.[73]

Prisoners

Over 10,000 officers and men were taken prisoner during the battle. The experience of the Gettysburg POWs is not our purpose here. We shall instead examine the process of how men were taken prisoner in the course of the battle. How easy was it for a combatant to surrender? Was it safer to surrender in groups rather than singly? Gettysburg occurred before the collapse of the prisoner exchange system, and although every soldier understood that prisons were notoriously unhealthy places to be avoided if possible, they had not yet reached the overcrowding that turned them into true hellholes. It was still believed that if you were captured, you had a reasonable chance of surviving your imprisonment.

Richard Holmes, in his book *Firing Line,* speculates that soldiers in twentieth-century wars who attempted to surrender after the fighting came to close quarters stood only a fifty-fifty chance of their surrender being accepted. This was certainly not true at Gettysburg. The soldier who surrendered in this battle, even in cases of close, or hand-to-hand, combat was nearly always spared. This is not to say that the killing of those attempting to surrender did not happen, as several accounts related below show, but it was rare. During the fighting along the Emmitsburg Road on July 2, Roland Bowen of the 15th Massachusetts continued to fire at the advancing Confederates of Wright's Georgia brigade, even though his comrades broke for the rear. Bowen resisted until it was too late to make a run for it without being shot down. So, he wrote, "I threw down my gun and held up both hands, my cap in one and begged that they might spare my life." Wright's men "spoke not a word to me but passed over and on." No one from Wright's line of battle even bothered to direct or escort Bowen to the rear. He did so on his own, since this was the only direction he could travel to find safety from the bullets and shell fragments flying about. Eventually he came upon a Confederate soldier, whom Bowen found "was mighty glad to get one prisoner to go the rear himself," who accompanied him to the rear.[74]

Many soldiers who surrendered recorded being threatened by their captors, or handled roughly initially, but in only the most exceptional cases were threats carried out. The experience of George W. Whipple of the 64th New York was fairly typical. During his regiment's disorganized retreat from an advanced position in the Wheatfield on July 2, Whipple's company commander, Captain Charlie Fuller, was badly wounded. Whipple tried to carry Fuller from the field, but was unable to keep up and had to lay him down. Moments after he did so the Confederates were upon him. Whipple wrote that they called out as they approached, "Surrender, you d-m Yankee." Whipple asked his captors if he could remain for a moment more with his captain, "but the bayonet was close to my back, with awful threats to put it through me if I refused." "Go to the rear you d——-d Yankee son of a b——-h," they cursed at him. Whipple complied and was escorted to the rear by two of his captors, who were probably very happy to have a reason to leave the front lines. They left him in the rear with some men from the Ambulance Corps, unharmed.[75]

Captain Alfred E. Lee of the 82nd Ohio, who was wounded on July 1 during the fighting north of Gettysburg, recalled that as the Confederate battle line overran his position, "Some of the victors seemed disposed to be savage. A wounded man lying near me, who had raised himself on his elbow, probably to get an easier posture, was assailed with a volley of curses by a stalwart soldier in gray, who ordered him to lie down instantly, on pain of being shot dead. The soldier held his musket at the ready, evidently intending to execute his threat if not summarily obeyed." The soldier complied and the threat to fire was not carried out. Moments later a Confederate skirmisher made his way through the dead and wounded of Lee's regiment. Lee asked him whether the Union wounded would be molested. The Confederate replied, "No; you need not be afraid. Ten minutes ago I would have shot you in a minute; but now that you are a prisoner you shall not be disturbed." He took Lee's sidearm and went on his way. Several minutes later a Confederate battery came up and unlimbered where Lee lay helpless. Some of the Confederate artillerymen noticed that Lee was in danger of being trampled upon by their horses and "two of them very gently removed me to a place of greater safety."[76]

One of the most remarkable incidents illustrating the restraint of the victors toward those they had vanquished involved the 6th Wisconsin Infantry on July 1. The 6th made a charge upon elements of Davis's Mississippi–North Carolina brigade that had taken shelter in the Railroad Cut during the morning fighting. Davis's men killed and wounded some 180

men in the 6th in the 175 paces it took the Wisconsin men to cover the
ground from the Chambersburg Pike to the cut, and continued to resist
until the 6th reached the edge of the cut. Lt. Colonel Rufus Dawes, commanding the 6th, who was coming up behind the main line of his regiment, heard shouts from up front. "Throw down your muskets! Down
with your muskets!" Running forward, Dawes discovered the cut below
him filled with "hundreds of rebels" still clutching their weapons. Dawes
demanded the surrender of the Confederates, and they complied. Recalling the incident years later Dawes wrote, "the coolness, self-possession,
and discipline which held back our men from pouring in a general volley
saved a hundred lives of the enemy, and as my mind goes back to the fearful excitement of the moment, I marvel at it."[77]

Dawes's regiment captured some 240 of Davis's men, whom a detach-
ment escorted to the rear. As they made their way over Seminary Ridge,
the prisoners passed the remnants of the 147th New York, which Davis's
Brigade had mauled minutes before the 6th turned the tables on them.
Some of Davis's men apparently taunted the soldiers of the 147th. A lieutenant in the New York regiment recalled that "the taunts of our prisoners only added to our bitterness and when the enemy had dropped their
guns our men also dropped theirs and went at them in a regular fist[i]cuff
scrimmage." Officers intervened and broke up the fight, and "the feeling
of anger soon passed away, forgotten in the joy that we were relieved and
our pursuers had become our prisoners."[78]

In another incident reflecting the concern about protecting prisoners
from further danger, Lieutenant Albert Clark of the 13th Vermont ordered the men of Pickett's Division who had surrendered to his company to lie down as the air "was thick with hissing shot." One of Clark's
men complained, "you are treating the enemy better than you treat us,"
to which Clark replied, "that is true, but we are at work and their work is
over."[79]

Not everyone who attempted to surrender was spared. In instances of
hand-to-hand combat, surrender might not be accepted because the individual attempting to surrender had crossed the line where because of his
actions he could not expect mercy from his foe. In the action at the Railroad Cut on July 1, before Davis's men surrendered to the 6th Wisconsin,
John Harland of Company I made for the flag of the 2nd Mississippi. A
Confederate soldier defending the flag shot and killed Harland, whose
body rolled down into the cut. Harland's comrade, Levi Teague, was right

behind him and he leveled his rifle at the Confederate who called out, "Don't shoot! Don't kill me!" An observer of the incident said Teague replied, "All hell can't save you now." He fired point-blank and killed the Confederate.[80]

It could also be risky to surrender while others continued to resist. This was true in the final moments of fighting at the Angle during the repulse of Pickett's Charge. Amos Plaistad of the 15th Massachusetts recalled in 1870 that when his regiment charged the stone wall Pickett's men had captured in their advance, the surviving Virginians attempted to run for it. Plaistad and his comrades fired a volley at them that caused the Confederates to drop down to the ground for cover, and they "cried out to us to stop firing and let them come in." The Federals did and some of Pickett's men started toward the Federals, but, wrote Plaistad, "the rest again started on the run towards the Emmitsburg Road," and his regiment opened fire on them. "We were obliged to fire through those who were ready to come in and many was killed coming tords [sic] us; but nearly all lay down until [sic] we ceased firing." The difference between this incident and that involving Lieutenant Clark is that Clark's prisoners were within his lines, while those attempting to surrender to Plaistad's regiment were caught between the lines, a perilous place for anyone attempting to surrender.[81]

Incidents of prisoners being shot after their surrender are conspicuously rare in the accounts of Gettysburg. In the only one discovered in the research for this paper, Edward Freeman of the 13th Vermont related that after the fighting on July 3, "my comrade shot a rebel right in the head because he would not give up his gun." Certainly there were other incidents of prisoners being shot after their surrender, or in the act of attempting to surrender, but the evidence indicates they were rare and isolated.[82]

The question these various stories raise is why did the captors spare those who moments before shot down their comrades? Why did the men of the 6th Wisconsin hold their fire when they could have taken revenge and massacred Davis's men? Why did the men of the 147th New York not simply shoot down some of the unarmed men of Davis's brigade who saw fit to taunt them, instead of dropping their weapons and going at them with their fists? Why were incidents such as the one Edward Freeman observed the exception rather than commonplace, as they would be in the twentieth century? This is a subject that deserves further study, but some reasonable speculations can be made. The combatants all possessed a common language, religion, and race—a commonality that encouraged soldiers to treat their captured enemy decently. Where race was an issue on other Civil War battlefields, brutal treatment and killing of prisoners was general. A sense of honor also restrained soldiers from shooting down unarmed or surrendering foes. To kill an unarmed man was seen as a cowardly act, even in situations where the surrendering enemy had killed or wounded many men in the victorious party, such as the 6th Wisconsin at the Railroad Cut. This also explains incidents like Lieutenant Clark's concern for the safety of the prisoners his company had captured. Clark saw it as his duty to protect these men who could no longer defend themselves.

Some men's hands may have been stayed by the fact that they were armed with single-shot muzzle-loading weapons. During the Civil War, it was not as easy to kill people as it would be when automatic weapons were introduced, where a squeeze of the trigger could gun down a group of men. Shooting prisoners also encouraged the enemy to fight more desperately, and soldiers understood that killing prisoners or men trying to surrender only encouraged retribution. The nature of Civil War combat, where infantry fought in linear lines, also made it easier—compared to twentieth century warfare—to distinguish those surrendering from those resisting. That this did not always ensure safety for those trying to surrender is born out by Amos Plaistad's statement. Finally, taking prisoners was a way to escape the firing line. This was the case with the Confederate who escorted Roland Bowen to the rear. And certainly George Whipple did not need two men to escort him to the rear. Even though men were not to leave the ranks for any reason unless ordered to do so by an officer, such incidents were not uncommon.

Carnage

"The field of battle is a sight never to be forgotten," wrote Sergeant George Bowen of the 12th New Jersey. Bowen, like nearly all the soldiers who fought at Gettysburg, was a veteran of earlier battles. He had seen the revolting aftermath of combat before and knew what to expect. But no one ever seemed to grow accustomed to the carnage of a battlefield, even though some soldiers claimed that they did. The scale of Gettysburg's slaughter was so vast that it surpassed anything veterans had seen before. Gettysburg was the first time that Bowen had an opportunity to walk a battlefield after a fight ended, since the Confederates had held the field in the previous battles he had participated in. "One time is enough," wrote Bowen; "the sight and the sound were terrible, no one can give an idea of what it is like, the pain and misery of those poor fellows whom we shot down only a few hours ago—it is a heartbreaking sight." Captain George K. Collins, a company commander in the 149th New York, walked the slopes of Culp's Hill on July 4 after the Confederates withdrew, to observe the effects of his regiment's musketry. "The impressions received during that morning walk will never be effaced from memory. It made the men sick both in body and mind. They had been nearly without food for forty-eight hours, yet it was with difficulty that they could retain food in their stomachs." Franklin Gaillard, the Lt. Colonel of the 2nd South Carolina, whose regiment had seen more than its share of hard fighting and bloody battlefields, wrote to his wife after Gettysburg, "It was the most shocking battle I have ever witnessed." Yet, Gaillard, Collins, Bowen, and thousands of other survivors remained in the ranks to do their duty, fully aware of the risks that this entailed. A sense of honor, dedication to their cause, and, importantly, a desire to never let their comrades down held men in the ranks in spite of the horrors they had witnessed and dangers they knew awaited them in the next battle. But the deep emotional scars the violence and carnage of a battle like Gettysburg left upon its survivors never truly healed. Colonel Eppa Hunton, who led the 8th Virginia of Pickett's Division at Gettysburg, was asked to return to the battlefield a number of times after the war. Hunton wrote, "I could never summon the courage to do so. If I were to go over the line of our charge I would say, 'Here fell Captain Green;' 'Here fell Captain Bissell;' 'Here fell Captain Grayson;' 'Here fell Captain Ayres'—& a host of others. It would nearly kill me to see where so many brave men fell—all of them among the best friends I ever had."[83]

Notes

1. John Keegan, *The Face of Battle* (London: Penguin, 1976).

2. William McCann, ed., *Ambrose Bierce's Civil War* (Washington, D.C.: Regnery Gateway, Inc., 1956; New York: Gramercy Publishing, 1996), 39. Citations are to the 1996 edition.

3. Abner Small, *The Road to Richmond* (Berkeley: Univ. of California at Berkeley, 1957), 184.

4. John M. Stone to General, n.d., Bachelder Papers, Gettysburg National Military Park Library [hereafter, "GNMP Library"]; Supreme Court of Pennsylvania, *Trial of the Survivors of the 72nd Pennsylvania versus the Gettysburg Battlefield Memorial Association* (1891), 83.

5. Christopher Smith, "Bloody Angle," Gettysburg Newspaper Clippings, vol. 4, 41–44, GNMP Library; John Gibbon, *Personal Recollections of the Civil War* (New York and London: G. P Putnam and Sons, 1928; Dayton, Ohio: Morningside, 1978), 147–148 (citations are to the 1978 edition); George K. Collins, *Memoirs of the 149th N.Y. Vol. Inft.* (Reprint, Hamilton, N.Y.: Edmonston Publishing, 1995), 139.

6. Robert L. Bee, ed., *The Boys from Rockville: Civil War Narratives of Sgt. Benjamin Hirst, Company D, 14th Connecticut Volunteers* (Nashville: University of Tennessee Press, 1998), 149.

7. Bernard Matthews, "Leg Taken Off by Gettysburg Shell," *The Louisville Evening Post* (July 3, 1913), in Gettysburg Newspaper Clippings, vol. 6, p. 102.

8. R. S. Robertson to parents, June 28, 1863, Vertical File V6-NY93, GNMP Library.

9. Charles T. Bowen diary, Vertical File V6-US12, GNMP Library.

10. James Houghton journal, 4th Michigan, Vertical File V6-MI4, GNMP Library.

11. William C. Oates, *The War Between the Union and the Confederacy* (Dayton, OH: Morningside, 1974), 212–222.

12. Ibid., 212.

13. Ralph O. Sturtevant, *Pictorial History: Thirteenth Regiment Vermont Volunteers, War of 1861–1865* (n.p., 1910), 323; Charles Haley journal, Vertical File V6-ME17, GNMP Library.

14. Michael Jacobs, "Meteorology of the Battle of Gettysburg," [Gettysburg] *Star and Sentinel,* August 11, 1885.

15. Anthony McDermott to John Bachelder, June 24, 1886, Bachelder Papers [hereafter, "BP"], GNMP Library; Sturtevant, 797; Joseph I. Durkin, ed., *John Dooley, Confederate Soldier His War Journal* (South Bend, IN: University of Notre Dame Press, 1963), 105.

16. Supreme Court of Pennsylvania, *Trial of the 72nd Pennsylvania,* 150; U.S. War Department, *War of the Rebellion: The Official Records of the Union and Confederate Armies* (Washington, D.C., 1889), vol. 27, pt. 1, 445–446, [These records from the War Department are hereafter referred to as "OR"]; letter from Captain George Hillyer, *Southern Banner,* July 29, 1863, V7-GA9, GNMP Library; Frank L. Byrne, ed., *Haskell of Gettysburg* (Madison: State Historical Society of Wisconsin, 1970), 150.

17. J. H. Moore, "The Battle of Gettysburg," *The Military Annals of Tennessee* (J. M. Lindsley & Co., 1886), copy Vertical File V7-TN7, GNMP Library; Henry T. Owen, "Pickett's Charge," *Gettysburg Compiler,* April 6, 1881; Byrne, ed., *Haskell of Gettysburg,* 150.

18. Small, *Road to Richmond,* 106; Gregory Acken, ed., *Inside the Army of the Potomac: The Civil War Experiences of Captain Francis Adams Donaldson* (Mechanics-

burg, Pa.: Stackpole Books, 1998), 305–306. Donaldson's letters truly rank among the classics of wartime writing from a combat soldier.

19. John Bigelow to John Bachelder, n.d., BP, GNMP Library; Survivors' Association, *History of the Corn Exchange Regiment 118th Pennsylvania Volunteers* (Philadelphia: J. L. Smith, 1888), 244; Franklin Whitmore to Parents, July 5, 1863, Vertical File (VF) 6-ME17, GNMP Library.

20. Collins, *Memoirs of the 149th*, 144–145.

21. William Hardee, *Instructions for Skirmishers* (New York: J. B. Lippincott, 1861).

22. Henry S. Stevens, Souvenir of Excursions to Battlefield by the Society of the Fourteenth Connecticut Regiment September 1891 (Washington, 1893), 11; Richard S. Thompson, "A Scrap of Gettysburg," Military Order of the Loyal Legion United States, Illinois Commandery, vol. 3 (Reprint, Wilmington, N.C.: Broadfoot Publishing, 1992), 98. Citations are to the 1992 edition.

23. Thompson, "A Scrap of Gettysburg," 98.

24. Noah Trudeau, ed., "5th Alabama Sharpshooters Taking Aim at Cemetery Hill: Eugene Blackford Memoir," *America's Civil War* (July 2001), 46–53.

25. William J. Seymour journal, James Schoff Collection, William L. Clements Library, Univ. of Michigan.

26. H. S. Stevens, Souvenir of Excursion to Battlefields, Society of the Fourteenth Connecticut Regiment (Washington, D.C.: Gibson Brothers, 1893), 16.

27. Ibid., 18.

28. George L. Yost to Father, July 4, 1863, 126th NY Regimental File, GNMP Library.

29. Mary Lasswell ed., *Rags and Hope: The Recollections of Val C. Giles, Four Years with the Texas Brigade* (New York: Coward McCann, 1961), 180.

30. J. B. Polley, *Hood's Texas Brigade* (Reprint, Dayton, OH: Morningside, 1976), 177. Citations are to the 1976 edition.

31. Gregory A. Coco, ed., *From Ball's Bluff to Gettysburg . . . And Beyond: The Civil War Letters of Private Roland E. Bowen 15th Massachusetts Infantry 1861–1864* (Gettysburg, Pa.: Thomas Publications, 1994), 201.

32. Ibid., 201.

33. Ibid.

34. The 15th Massachusetts lost eleven officers killed and wounded on July 2 and July 3, which indicates that many officers were doing their duty during the fighting. Possibly, the lack of leadership only existed in Bowen's company.

35. Acken, ed. *Inside the Army of the Potomac*, 303.

36. John C. Reed diary, Alabama State Archives. Reid's "diary" is in fact his recollections.

37. Joshua L. Chamberlain to General Barnes, Sept. 3, 1863, New York Historical Society (typescript copy GNMP Library).

38. Charles A. Fuller, *Personal Recollections of the War of 1861* (Reprint, Hamilton, N.Y.: Edmonston Publishing, 1990), 94–95.

39. *The Cattaraugus* [N.Y.] *Freeman*, July 8, 1863. This is an account of the 64th

New York's action at Gettysburg by their colonel. Of the two color bearers, one was killed and the other wounded before the charge ended.

40. For details on the battle between these two regiments see, Rod Gragg, *Covered With Glory: The 26th North Carolina Infantry at the Battle of Gettysburg* (New York: Harper Collins, 2000), and O. B. Curtis, *History of the Twenty-Fourth Michigan of the Iron Brigade* (Detroit: Wing & Hammond, 1891; Gaithersburg, Md.: Old Soldier Books, 1989). Citations are to the 1989 edition.

41. Charles D. Page, *History of the Fourteenth Regiment Connecticut Volunteer Infantry* (Reprint, Gaithersburg, Md.: Ron R. Van Sickle Military Books, 1987), 152.

42. Collins, *Memoirs of the 149th,* 143.

43. Ibid., 144; Robert G. Scott, ed., *Fallen Leaves: The Civil War Letters of Major Henry Livermore Abbott* (Kent, OH: Kent State Univ. Press, 1991), 184.

44. OR, vol. 27, pt. 1, 439.

45. Scott, ed., *Fallen Leaves,* 188.

46. Franklin Gaillard to Maria, July 17, 1863, Gaillard Papers, Southern Historical Collection, University of North Carolina at Chapel Hill [hereafter abbreviated as SHC, UNC]; Joseph B. Kershaw, "Kershaw's Brigade at Gettysburg," *Battles and Leaders of the Civil War* (New York: Century, 1884–1888), vol. 3, 336.

47. OR, vol. 27, pt. 1, 624.

48. Oates, *The War Between the Union and the Confederacy,* 219.

49. Thomas A. Desjardin, *Stand Firm Ye Boys from Maine* (Gettysburg, Pa.: Thomas Publications, 1995), 74.

50. For details about this action, see D. Scott Hartwig, "It Struck Horror to Us All," *Gettysburg Magazine,* No. 4 (January 1991), 89–100.

51. "Report of reunion of the officers of the Army of the Potomac, August 23–28, 1869," Typescript, p. 25, Huntington Library, San Marino, Calif.; John S. Patton Papers, Historical Society of Western Pennsylvania Collection.

52. James Houghton journal, Bentley Historical Library, University of Michigan; Robert H. Campbell Reminiscence, Bentley Historical Library.

53. John Reynolds, "The Nineteenth Massachusetts at Gettysburg, July 2–3–4, 1863," VF6-MA19, GNMP Library; "Ben Hirst's Narrative," in Gary W. Gallagher, ed., *The Third Day at Gettysburg and Beyond* (Chapel Hill: University of North Carolina Press, 1994), 140–141; Edward P. Alexander wrote that a gun in action "can easily fire from 30 carefully aimed shots in an hour, to 100 hurriedly aimed." See, Gary W. Gallagher, ed., *Fighting for the Confederacy: The Personal Recollections of General Edward Porter Alexander* (Chapel Hill: University of North Carolina Press, 1989), 246.

54. John H. Rhodes, *The History of Battery B, 1st Rhode Island Light Artillery* (Providence, R.I.: Snow & Farnham Printers, 1894), 209–210.

55. Christopher Smith, "Bloody Angle." Gettysburg Newspaper Clippings, vol. 4, 41-44, GNMP Library; John H. Reynolds, "The Nineteenth Massachusetts," Typescript, VF6-MA19, GNMP Library.

56. Christopher Smith, "Bloody Angle." Smith's memory clearly failed him several times when he wrote this article. He remembered Sgt. Whitston as Sgt. Watson, and Arsenal Griffin as William Griffin. He also described Griffin as a teamster. It is possible he served on the commissary wagon, but he was more likely a limber driver. The correct names for these men is found in the June 30 muster report for Battery A, 4th U. S. Artillery in the National Archives. A typescript copy of this muster roll is available at the GNMP Library.

57. Ibid.; Edward P. Alexander, *Military Memoirs of a Confederate* (New York: Charles Scribner's Sons, 1907; Dayton, OH: Morningside Bookshop, 1977), 421. Citations are to the Morningside edition.

58. OR, vol. 27, pt. 1, 432; Kent M. Brown, *Cushing of Gettysburg: The Story of a Union Artillery Commander* (Lexington: University Press of Kentucky, 1993), 258, 264–267.

59. Scott, ed., *Fallen Leaves,* 186.

60. James Wright Reminiscence, Minnesota Historical Society, VF6-MN1, GNMP Library; Anthony McDermott to John Bachelder, June 2, 1886, John Bachelder Papers, New Hampshire Historical Society, copy GNMP Library; John D. Smith, *History of the Nineteenth Regiment of Maine Volunteer Infantry 1862–1865* (Minneapolis: 1909; Gaithersburg, Md.: Ron R. Van Sickle, 1988), 99. Citations are to the 1988 edition.

61. Polley, *Hood's Texas Brigade,* 167; G. W. Finley, "The Confederate's Story," *Buffalo Evening News,* May 29, 1894, copy Vertical File V7-VA56, GNMP Library; Franklin Gaillard to Maria, July 17, 1863, Gaillard Papers, SHC, UNC; Scott, ed., *Fallen Leaves,* 188.

62. William J. Seymour Memoir, William L. Clements Library, University of Michigan; OR, vol. 27, pt. 2, 480.

63. OR, vol. 27, pt. 1, 886; Rhodes, *The History of Battery B,* 202.

64. Strength and losses are drawn from John Busey and David Martin, *Regimental Strengths and Losses at Gettysburg* (Hightstown, N.J.: Longstreet House, 1986).

65. Birkett Fry to John Bachelder, Jan. 26, 1878, in David and Audrey Ladd, ed., *The Bachelder Papers* (Dayton, Ohio: Morningside, 1994), vol. 1, 522.

66. E. T. Boland, "Beginning of the Battle of Gettysburg," *Confederate Veteran,* vol. 14, 308; A. S. Van de Graaf to Wife, July 8, 1863, Vertical File V6- 5AL battalion, GNMP Library.

67. Busey and Martin, *Regimental Strengths,* 292. Curiously, few of Buford's men, or writers who attempted to make Buford's morning stand against Heth the stuff of legends, gave much attention to this very significant afternoon action where his troopers did indeed slug it out with Confederate infantry and inflict heavy losses.

68. Thomas L. McCarty, "Address," Robert Brake Collection, United States Military History Institute.

69. Untitled manuscript, Henry Clay Potter Papers, Gettysburg College, quoted in Paul M. Shevchuck, "The 1st Texas Infantry and the Repulse of Farnsworth's Charge," *Gettysburg Magazine,* No. 2 (January 1990), 86.

70. Jeffery Stocker, ed., *From Huntsville to Appomattox* (Nashville: University of Tennessee, 1996), 111.

71. George W. Beale, *History of the Ninth Virginia Cavalry in the War Between the States* (Richmond, Va.: B. F. Johnson Publishing Co., 1899), 87–88.

72. Eric J. Wittenberg, ed., *At Custer's Side: The Civil War Writings of James Harvey Kidd* (Kent, Ohio: Kent State University Press, 2001), 23.

73. It should also be noted that much of the force of Hampton's and Lee's attack was checked by a heavy fire from Union horse artillery. The largest collection of personal accounts of the East Cavalry Battle can be found in *The Bachelder Papers*, volumes 2 and 3.

74. Richard Holmes, *Firing Line* (London: Jonathan Cape, 1985), 383; Coco, ed., *From Ball's Bluff*, 203.

75. Letter of George W. Whipple printed in the *Cattaraugus Freeman*, January 9, 1864.

76. "Reminiscences of the Gettysburg Battle," *Lippincott's Magazine*, vol. 6 (1883), 57. The author of this piece identified himself as "a company officer," but research has established that this was Alfred E Lee, of Company E.

77. Rufus R. Dawes, *Service with the Sixth Wisconsin Infantry* (Madison: Wisconsin State Historical Society, 1962), 168–169.

78. James Coey, "Cutler's Brigade—Magnificent Fight on the First Day at Gettysburg," Vertical File V6-NY147, GNMP Library.

79. Sturtevant, *Pictorial History*, 593.

80. Lance J. Herdegan and William J. K. Beaudot, *In the Bloody Railroad Cut at Gettysburg* (Dayton, Ohio: Morningside, 1990), 200.

81. Amos Plaistad to John Bachelder, June 11, 1870, in Ladd, ed., *The Bachelder Papers*, vol. 1, 393.

82. Sturtevant, *Pictorial History*, 528.

83. "The Diary of Captain George D. Bowen," *Valley Forge Journal*, vol. 2, no. 1 (June 1984), 134; Collins, *Memoirs of the 149th*, 147; Gaillard to Maria, July 17, 1863; Eppa Hunton, *The Autobiography of Eppa Hunton* (Richmond, Va.: William Byrd Press, 1933), 100.

3
Effects of Battle

Wounds, Death, and Medical Care in the Civil War

Bruce A. Evans, M.D.

Oliver Wendell Holmes, Jr., from a Memorial Day vantage point more than thirty years removed from the actuality of his war, made the well-known statement: "Through our great good fortune, in our youth our hearts were touched with fire."[1] Holmes himself was directly touched three times, including a wound sustained at Ball's Bluff that nearly did reach his heart. His personal experiences and the growing carnage around him finally led him to leave the war when his unit's enlistment ended in 1864. The perspective of years and the softening of memories allowed him to recover some of the positive meaning of his history.

Like the young Holmes, the new and untouched soldiers rallying to the flag in 1861 looked towards the panoply of war with confidence and a sense of adventure, sustained by an ignorance of the brutal experience to come and a strangely bloodless traditional view of martial glory. The "empty chair" was more romantic concept than reality. Even after three hard years of war, when only the brave, the dogged, and the unimaginative were left on the battle lines, a sense of personal immunity may have sustained the ability of the soldier to face the best efforts of the enemy to do him a very personal harm.

To about a sixth of those serving three years,[2] the unreality of a personal exemption from the surrounding "stifled crash of balls hitting bones, and the soft chunk of flesh wounds mingled with the outcries of sufferers"[3] was abruptly announced by their wound. This caused each to cry "out when he was hit, uttering either an oath, or the simple exclamation 'Oh!' in a tone of dolorous surprise."[4] The eye of history generally remains focused on the unhit. Here, we will examine the story of the human wreckage left behind, and those who struggled to aid them.

The essence of nineteenth-century warfare was to kill, disable, or intimidate the enemy by propelling pieces of metal of various sizes in his direction by various means. Almost 400,000 such collisions were documented in the Union Army.[5] The immediate effects of the collision of the human body with these projectiles were various. Large high-velocity projectiles such as round shot or large exploding projectile fragments would meet little resistance from the frail human body, resulting in large parts being simply carried away. Striking the soldier once they had lost most of their velocity or obliquely, their considerable momentum nevertheless could wreak awful damage, including limbs shattered by round shot and deeply contused, lacerated, and with jagged wounds from shell fragments. Deceptive wounds could be delivered by nearly spent artillery projectiles, with what appeared to be merely a bruise concealing disruption and rupture of internal organs, such as intestines or spinal cord, or extensively shattered bone. These missiles could also play grim jokes, such as the fatally wounded soldier found to have an entire 12-pound round shot concealed within his buttocks.[6]

The bulk of the damage, however, was done by small arms fire. While at close range the damage done by a ball and a conoidal rifle projectile (minie "ball") was similar, the ability of the latter to hold its velocity over a longer range and the wedgelike profile contributed to the minie ball's greater destructive potential. The conoidal "ball" was less likely to be deflected, and more likely to pulverize and shatter bone. At the same time, the soft lead of the projective was liable to sometimes bizarre deformations—these were ugly, lacerated wounds, with the soft, deforming lead often pushing a plug of dirty frock coat, grimy shirt, and underclothes deep into the wound. These projectiles also could play hide and seek—the final resting place of the slug might be quite different from that expected from the entrance wound. The bullet wounding one soldier beneath the right jaw traveled to the left, down his neck, and then into his left shoulder blade.[7]

Approximately 17 percent of the soldiers hit on the field of battle died where they were hit. For the fortunate, death was swift: head wounds, heart wounds, or tearing damage to the largest blood vessels such as the aorta tended to be merciful killers. Surgeon John R. Lidell examined 43 dead following the attack and defense of Fort Steadman on the Petersburg lines. He found 23 head wounds, 15 wounds in the chest, and 5 abdominal wounds. Nearly all the dead with chest and abdominal wounds were pale, and with the chest wounds large amounts of blood were often evident in the clothes and on the ground.[8]

Just as quick but harder to conceive were the results of massive and overwhelming trauma. Every dramatic story of "double canister at point blank range" implies those on the receiving end essentially vaporized as they encountered a concentrated and expanding cone of forty-four 1.5-ounce lead balls. Solid shot, and to a somewhat lesser extent case shot, could and frequently did suddenly mangle a soldier beyond recognition. It is frequently cited that 95 percent of all reported wounds were caused by bullets,[9] but we should remember that many of the encounters with artillery projectiles did not allow the victim the luxury of wounded status—they were often killed where they stood.

For those not killed immediately, the recognized initial responses of the body to these insults are "shock" and pain. The symptoms of shock are paleness, trembling, coolness, and a diffuse perspiration, accompanied by a "feeble circulation."[10] These symptoms were felt to represent a "severe perturbation of the nervous system," proportionate, in general, to the severity of the wound.[11] The knowledge that these symptoms were also due to loss of circulating blood volume lay in the future. Fortunately, the old practice of a depleting treatment of therapeutic bleeding to treat the obvious inflammation associated with all wounds had been long banished due to negative effects obvious to all not blinded by theory.

The other immediate symptom was pain, ranging from "stinging" pain like a "hot wire" and pain that felt "like a blow"[11]—accompanying flesh wounds—to severe, all-encompassing pain accompanying nerve wounds or shot fractures. With massive trauma, a period of relative freedom from pain might initially be present, but if the wounded soldier survived more than a few minutes, pain became his steadfast companion on the battlefield.

Once wounded on the field, access to aid was not necessarily immediate. For those falling in disputed territory, a terrifying, lonely, and painful period of abandonment followed the wound. In extreme cases, such as occurred following the ultimately futile Federal attacks at Cold Harbor, the wounded lay unaided with the dead for days, with many joining their former comrades as aftereffects of the wounds took their toll.

If not immediately killed or rendered unconscious, the most immediate concern of the wounded was to discover the seriousness of their wound, especially during the initial period of relatively painless shock and the accompanying adrenaline rush. Frantic removal or pushing aside of clothing to visualize the wound was common, both to overcome the initial disbelief and to gauge the chances for survival. As time passed, pain and a ferocious thirst triggered by blood loss and worsened by the heat of many summer battlefields became the definition of the wounded soldier's existence. The moans and cries of pain mixing with the tortured pleas for water of hundreds of men seared the memories of unwounded listeners.

Delayed death was most likely to be due to gradual blood loss, complicated by increased fluid loss due to stifling heat and direct sun on many battlefields, or by freezing cold leading to hypothermia as at Fredericksburg or Fort Donaldson.

Even the dead refused to abide by the innocent romantic conceptions of warfare. Heat combined with bacterial production of gas in the gastrointestinal tract led to grotesque swelling of the bodies in a surprisingly short time. The effect of direct sunlight on unprotected and unmoving bodies produced a dramatic blackening of the skin that when combined with bloating turned even the soldier falling quickly from a merciful bullet wound into an object of horror within hours. The appearance of these unexpected images set before the public, as in the Antietam series of photographs presented by Gardner, must have been a shock indeed.

If not killed outright or abandoned between the lines, the wounded soldier would come to or be delivered into the hands of an army surgeon. The background, training, knowledge, and skills of that surgeon and his colleagues became at that moment the most important determinant of that soldier's fate.

As a result of the egalitarianism of the Andrew Jackson era, medical licensure with specific educational requirements was a thing of the past in the few states in which it had ever been established. The presence of numerous, active, and popular medical sects and cults prevented orthodox medical societies from exerting the type of effective control over a would-be physician's ability to practice that was present, for example, in England. For the medical consumer, it was truly a time of *caveat emptor*. Traditionally trained "allopathic" physicians attended two nine-month sessions of medical classes to earn a diploma, the second of which was a repeat of the first. Clinical training with supervised exposure to patients was available in few schools and required for graduation in fewer. The best physicians supplemented this training with experience in the renowned clinical medicine centers of Edinburgh, London, Paris, and Vienna.

At the start of the war, the U.S. Army Medical Service consisted of 114 regular army surgeons and assistant surgeons, of whom 24 resigned to go south and 3 were dismissed for disloyalty.[12] Even this small core had experience mostly with managing the health of companies or platoons on the frontier. From this minimal beginning, by April of 1865 more than 12,000 physicians had seen service in the field with regiments as assistant surgeons or surgeons, brigade surgeons, and corps or army staff medical officers, or in the general hospital system as army surgeons or civilian contract physicians.[13] The manpower challenges facing the South were, if anything, even worse.

The obstacles to an effective medical service were immense. Similar to Lincoln's characterization of the armies about to face off across Bull Run, these medical forces were "all green together." To this inexperience was added the effects of the deficiencies of the medical education system, the enormous deficit in the supply of physicians with significant surgical experience, and the political appointment of physicians to state volunteer regiments (including the odd homeopath or folk practitioner quite free of any formal training). Nevertheless, by 1864 experience had been gained, lessons largely learned, and the incompetents weeded out by examining boards.[14] The medical force charged with the care of the wounded and sick was generally as competent as the medical times and the scope of the demands would allow.

The medical care that was available to the wounded soldier during the Civil War is often viewed in retrospect as the tail end of the medical dark ages, with the potentially lifesaving advances to come tragically just out of reach. From that viewpoint it is a small step to convert ignorance of the future—a universal human condition—to ineptitude and a tendency to "butchery" that would be almost comic if it had not been so appalling in its effects. To evaluate the battlefield and hospital medical care of this war, it is essential to realize what knowledge and tools were available to bring to the task at that time, and equally essential to eschew the arrogant habit of measuring their merit and effectiveness on a modern scale. Not to do so is to demean the struggles of the army medical personnel to mitigate the brutal effects of the nineteenth-century warfare and to trivialize the suffering of the killed and wounded with an added pathos which their very real sacrifices do not require.

In 1861 medical knowledge was beginning to evolve from a mixture of theoretical "systems" of disease causality dictating theoretical treatments—often nonsensical to modern eyes—and a more practical empiricism that was somewhat less respectable academically. Lacking specific knowledge of most physiologic processes or of any specific disease cause, physicians focused on symptoms in their conception and treatment of disease. The most prevalent theoretical systems in the United States contributed a number of "stimulating" and "depleting" treatments, the use of which was selected based on an estimation of which type of treatment was likely to revert the constitution of the patient towards a normal state. A patient with a condition characterized by physical collapse, a weak, thready pulse, a "low, muttering" delirium, and lacking physical evidence of the red, angry swelling of inflammation, was treated with stimulants such as alcohol to increase the excitement of the system and tonics such as iron compounds to build it up. On the other hand, obvious inflammation, a high, "hectic" fever, a fast, pounding pulse, and a manic, raving delirium were among the signs of excess excitement of the patient's system and dictated the use of one of the "depleting" therapies meant to reduce inflammation and return the system towards normal. Depletion ranged from the bleeding and "counter-stimulating" blistering of the skin to medication-induced vomiting, sweating, and purging. Balanced against these theoretical treatments was simple experience. Some of this added tools to the theory-based treatments, including effective pain medicines such as morphine derivatives, anesthesia, smallpox vaccination, vegetables for scurvy, and quinine for "intermittent fever." Stark experience eventually tempered many of the excesses of the theoretical treatments such as therapeutic bleeding, which was virtually abandoned. The medical outlook of most well-trained physicians at this point was a product of the tension among these elements.

Advocates of experimentally based scientific medicine, led by the "numeric school" of Paris, were challenging this status quo. Specifically, many "heroic" depleting treatments such as violent purging were rejected. While both intellectually attractive and eventually productive, this approach had no actual improved treatments to offer at this time. Being seen therefore as essentially nihilistic by most practicing physicians who, after all, had patients who needed *some* treatment, the practical effects of the "numeric school" were limited to an erosion of the worst excesses of the theoretically based treatments.

A wounded soldier who could move on his own, or who was accessible to help, would initially be treated at the regimental aid station. This would be an area in immediate proximity to the battle line, chosen both for ease of access and, where possible, for shelter from direct fire. The regimental assistant surgeon, assisted by an aide, evaluated each injury. The equipment of the aid station was limited to the surgeon's pocket kit (a small, portable selection of instruments) and a field medical kit, containing a limited number of medications and bandaging supplies. Triage was performed, and obviously fatally wounded soldiers with missile wounds to the brain or abdomen were given opium or morphine for pain, while transport was reserved initially for those who could be saved. For others, treatment consisted of pain relief, stimulants against the depressant effects of the injury and blood loss (almost exclusively distilled spirits), and temporary control of bleeding through the use of tourniquets or pressure. Rough splints were applied to shot fractures. Quick dressings covered wounds. No definitive surgical treatment was undertaken at the aid station, but was deferred to the division and corps hospitals, where the most effective operating surgeons from the constituent regiments were gathered at a preselected site with full surgical kits and hospital supplies.

Evacuation to the more remote hospital site was the job of the ambulance service. Early in the war this was a haphazard process involving whatever band members or other soldiers were detailed by the colonel for the duty. Those chosen were usually the shirkers, the cowards, or the hopelessly incompetent, who were of no help on the battle line. Ambulances were under the command of the quartermaster department, and had often been sent on nonmedical errands by the line officers and were unavailable. By 1864, professional detachments of ambulance drivers and aides, well drilled and under direction of the medical officers, were available to perform the evacuations.[15] Under any circumstances, the trip was a horrific one for many wounded soldiers, with wagons traveling over rutted or muddy roads, jarring splintered limb bones and swollen, painful wounds.

Experience demonstrated that consolidating the main surgical effort of a division or corps into a single field location was preferable to uneven performance and supply of many isolated regimental field hospitals. Surgery was then put in the hands of the best operators among the regimental surgeons, with the others assisting at surgery and performing dressing and medication duties. The surgery needed was best performed in the field; despite the primitive conditions compared to fixed general hospitals far from the battlefield, the wounded soldier had the best chance of recovery, and of survival, when definitive surgery was performed within twenty-four hours of the wounding. Given the state of transport available, this usually meant the wounded soldier faced his trial in the field.

The corps or army medical director chose the location prior to the battle. The main considerations were road access, a level and well-drained area, the availability of a water supply, and probable remoteness from active fighting.

The Federal field hospital often consisted of a group of division hospitals at a single location, constituting a corps hospital. Earlier in the war, barns, mills, schools, and churches were used for the dispersed field hospitals. Later there was an increasing use of tent hospitals for the sake of light, ventilation, ease of isolation of contagious cases, and because—for reasons mysterious to the surgeons—there was often a lower incidence of certain complications, such as tetanus, in tents than in barns.

The object of the surgeons' attentions to the wounded soldier was to prevent immediate effects of the wound from killing his patient and to encourage the wound to heal. This problem in the individual soldier could be approached through the principle of "conservation"—minimal operative interference and dependence on the restorative powers of the body—or through active surgical effort.

All wounds were observed to undergo a process of "reaction," during which they would become swollen, red, and hot. This process would begin within minutes to hours, and become well established after twenty-four hours, often accompanied by prostration and "primary wound fever." Early on, the wound would discharge a watery fluid. Sometime after the establishment of the reaction, a wound would begin to discharge pus. This was so invariably present that it was assumed to be an integral part of the healing process. The best result was the gradual filling up of the wound from inside out with healthy pink scar tissue, a gradual reduction in the pus discharge, and resolution of the swelling, heat, and accompanying fever of the reaction process. If things did not go well, the discharge remained copious or increased, the tissue reaction continued or progressed, and the patient became prostrate or moribund. The end could be quiet as the patient simply sank into coma and death, or come dramatically in minutes with erosion of arteries causing a fatal "secondary" hemorrhage, or in hours with the appearance of a recognized disorder such as blood poisoning (pyemia—literally "pus in the blood"). From the first medical contact, the issue for the Civil War surgeon was to decide on the degree of intervention that would maximize the chances of life.

Wounds to the brain, chest, abdomen, and spine were usually fatal, regardless of treatment. Other wounds, such as soft tissue or extremity wounds, would often heal. Some things were known to lessen the chance of the healing process progressing satisfactorily. Firstly, foreign bodies in the wound resulted in a non-healing wound with constant pus discharge. This was invariable with pieces of clothing and other incidental passengers of the bullet. The body might, however, tolerate the presence of the bullet itself, healing after "encysting" the missile. Secondly, pieces of bone separated by the impact, or badly bruised, usually died and functioned as foreign bodies ("sequestrum"), preventing healing. Finally, extensive blood vessel and tissue destruction simply asked more of the body's healing powers than it could supply.

Patients undergoing surgical procedures intended to increase the chances of a good outcome by removing these impediments to healing nevertheless could and did die. The reason was not clear. The shock of surgery on a failing constitution was often cited. The fact that from a modern perspective the surgeries introduced and spread infection in direct proportion to the aggressiveness and scope of the procedures was unsuspected, as was the role of blood loss and failure of the circulation. From educational and personal experience, each surgeon had to learn where the balance between help and hurt lay.

Although antisepsis lay unseen in the future, surgical procedures had lost some of their horror with the widespread availability of anesthesia. Dramatic drawings of struggling wide-eyed patients and claims for bullet relics with tooth marks notwithstanding, fewer than 300 surgical procedures were not performed under anesthesia (ether or chloroform) in the Union Army.[16] This was also true in Confederate hospitals, despite the lack of indigenous ability to manufacture these substances in adequate amounts. In fact, in addition to surgery, anesthesia was often used for the application of painful remedies, or thorough examination of wounds. Few deaths from anesthesia were reported.

Upon arrival at a field hospital, the first responsibility of the assisting surgeon was to thoroughly examine the wound. The purpose was to guide the decision as to the most appropriate treatment. Is the wound clearly mortal? If not, are there factors present that suggest that a surgical procedure will increase the chance that the patient will live? If not, does the wound need local treatment such as probing for foreign bodies whose removal will increase the likelihood of uneventful healing? The answers to these questions required knowledge of the path and effect of the bullet. Probing of the wound was therefore carried out, usually under anesthesia. Where did the bullet go? Into muscle but not into the abdomen—good, I'll probe for foreign bodies and clean the wound. Into the viscera— nothing I can do. Did that bullet splinter the bone? Yes, shot fracture— can I get the all fragments out? Adequate examination might require enlargement of the wound with a surgical knife. The most "educated probe" was the finger.[17] Despite a modern shudder, based on the knowledge of the time this made perfect sense: you can "feel" with a finger, and metal probes often did extensive tissue damage that a finger would not do.

Soft tissue wounds of the extremities sparing the bones and joints rarely carried a risk of death necessitating amputation. Exceptions included extensive soft tissue damage rendering the limb more burden than asset, or damage to major arteries of an extent preventing an adequate blood supply to the limb. Over 50,000 soft tissue injuries to the lower extremity were reported to the Union Medical Department during the war. Only 201 of these resulted in amputation (most amputations were done for wounds associated with bone fractures), and many of these were secondary procedures performed after the development of hospital gangrene in the limb.[18] The other main cause of amputation was destruction of major arteries by the original injury.

For most of these wounds, once they had been examined and probed and any foreign bodies removed, expectant treatment was the rule. The mortality of flesh wounds varied by area, ranging from 14 percent in the neck to 1 percent in the chest.[19]

Penetrating wounds of the brain, lungs, and abdomen were frequently fatal. What is more, experience had adequately shown that the chances of survival were not improved by surgical treatment beyond the examination necessary to establish the facts. In these cases of usually fatal wounds, the course of conservation was dictated mainly by frustration.

Wounds of the head directly penetrating the brain through the tough fibrous covering of the dura mater were fatal. A soldier with such a wound who survived long enough to arrive at the field hospital succumbed within hours, or a day or two at the most. From a modern perspective, immediate death resulted from damage to portions of the brain controlling vital bodily functions, or from massive bleeding causing secondary pressure on those structures. Delayed death resulted either from swelling of the brain tissue associated with the injury causing similar pressure, or from infection caused by the intrusion of the missile and associated external matter. Whatever the cause, remedy was beyond the knowledge and tools of the time. Any treatment, including probing for and removal of the missile and any foreign matter propelled into the wound by its violent entrance, were futile, as additional deterioration from swelling and further cause of infection only hastened the end. The pitiful victims of such wounds, consequently, were usually kept as comfortable as possible, as the inevitable end ensued. Occasionally, however, a surgeon could not keep himself from attempting to intervene. Union millionaire Brigadier James Wadsworth was wounded near midday during Longstreet's flank attack in the Wilderness when his horse bolted towards the enemy. A bullet entered the back of Wadsworth's head, lodging in the left frontal lobe, "splattering brains over his aide's coat."[20] Later that evening in a Confederate hospital, insensitive, with his "mouth . . . drawn down on the left and his right arm . . . paralyzed," his stare was vacant. He died that evening despite (or perhaps hastened by) an operative attempt to probe for and remove the bullet—an attempt felt to be misguided by an observing Federal surgeon.[21]

The surgeons were in a similar position of impotence when faced with soldiers who had penetrating bullet wounds of the abdomen. From our perspective, infection of the abdomen—peritonitis—is almost inevitable in these circumstances and almost invariably fatal if no intervention is undertaken. Once again, however, without antisepsis any effort of the contemporary surgeon only made things worse. Unlike with the brain wounds, however, these patients died a painful lingering death, conscious to near the end. The speed at which the end was reached depended mainly on the extent of bowel injury and the consequent fecal soiling of the abdomen. Lt. Gen. J. E. B. Stuart died "in his own blood and feces" in a bed in Richmond an agonizing twenty-seven hours following his wounding at Yellow Tavern. Wounded by a pistol bullet striking him below the ribs on the left side, he was able to reminisce, and in the style of the times consign himself to God's will and even attempt to participate in the singing of hymns. The inevitable end was known to all, including Stuart, from the start.[22]

On the other hand, apparently fatal wounds such as the pelvic wound received by Brig. Gen. Joshua L. Chamberlain at Petersburg the same year, were occasionally survived, sometimes through a daring and creative surgical effort. Friends of Chamberlain resisted his triage to palliative care and expected death. Two regimental surgeons, Shaw and Townsend, reconstructed his torn urethra and reattached it to his damaged bladder, working around a silver catheter inserted through the urethra into the bladder. His life was saved, although the rest of his long life was plagued by urine leakage through the pelvic floor and energy-draining infections.[23]

Penetrating chest wounds, while not invariably fatal, did not seem to benefit from any particular treatment or surgical intervention. Some patients survived, usually with slowly healing wounds draining infected matter. Many died. Spurred by the bubbling of air through the wounds and the obvious respiratory distress of many patients, many surgeons experimented with sealing the wounds with airtight materials such as rubberized dressings and collodium (gun cotton) glue. Although this often benefited the respiratory distress, the patients were more likely to finally succumb.[24] From a modern perspective, without an external drainage site for infected matter, internal infection eventually took the patient's life, occasionally after a prolonged and enervating struggle.

These fatal wounds, for which contemporary surgical treatment had generally been found at best ineffective and often harmful, have several factors in common from a modern perspective. In some cases, death resulted directly from damage to indispensable structures (the brain, large arteries of the chest, abdomen, or pelvis) which were not amenable to repair with the medical technology of the times. In other cases the wounds involved violation of sterile body cavities, with inevitable infection only worsened by additional surgical violation. Death resulted in these cases from the effects of that infection, either through a slow wearing down of the patient, involvement of vital structures, or the spreading to the blood of the fatal pyemia. Survival, while unlikely, occurred when the body was able to wall off the infection as an abscess. Also required for survival was a route for drainage of infected and inflammatory material to the outside, allowing the wound to slowly heal from the inside out. Sometimes such survival meant living with a chronic draining wound for the rest of the patient's life. The surgeon's role in these cases was most often a simple acceptance of his limitations.

With wounds of tissues more tolerant of surgical manipulation or sacrifice, the surgeon of 1864 did have the opportunity to save lives through direct effort.

Wounds directly damaging arteries threatened life through primary hemorrhage. Control of the bleeding by tourniquet or pressure might allow a soldier to reach the field hospital, but more permanent treatment would be required for survival. The usual treatment was an operative approach to tie off (ligate) the artery in the wound, or above the wound. Some wounds were fatal because of damage to an artery that could not be ligated, either because of a deep or otherwise inaccessible location, or because the blood supply of the artery was necessary for other vital structures. Some patients died because of shock (circulatory collapse) due to cumulative loss of blood volume despite eventual control of hemorrhage. Many died with recurrent hemorrhage, either immediately due to inadequate control of the damaged artery, or later—secondary hemorrhage—as the ligated artery sloughed away during the "healing" process.[25] Although victims were treated with various stimulants and tonics, there was no way to increase the blood volume directly, and patients died who in later years would have been saved with blood transfusions or intravenous fluids. Interestingly, it had been known in India that patients dying of cholera (who essentially die of shock due to fluid loss from the bowels) could be almost miraculously revived from the edge of death by introduction of salt fluids into the vein. However, they would then die within twenty-four hours of pyemia. This fate is unsurprising given the nonsterile nature of the fluids introduced. Direct treatment of shock therefore remained out of reach.

When a wound involved injury to bone—most often a "shot fracture"—the mortality rate immediately increased compared to an injury of the soft tissues alone. Fractures associated with penetrating gunshot wounds are by definition compounded fractures, defined as a fracture exposed to the air through a skin defect. In the first two thirds of the nineteenth century, compound fractures were deadly wounds.

With a few exceptions, shot fractures of the skull were treated conservatively. Only if a piece of bone from the inner table of the skull was felt to be pressing into the brain in a deteriorating patient, or if a fracture lay over the course of a major skull artery in a patient showing signs of deterioration from bleeding on the brain, was any surgery attempted. The reluctance to stray from the course of conservation stemmed from the poor results of either simple exploration with removal of fragments or trephination under battlefield conditions. In trephination, a disk of skull is removed by a circular saw apparatus to allow access to the space between the skull and the dura mater, a tough lining around the brain.

With that access, adjacent depressed fractures could be elevated, comminuted fragments of bone removed, bleeding arteries underlying the skull ligated, and (rarely) hemorrhage between the skull and the dura drained. The results were dismal, with 45 percent of the patients dying, and with even higher mortality when performed shortly after the injury in field hospitals.[26]

No medical procedure more typifies the general conception of the physical consequences of Civil War battle than amputation. Contemporary descriptions of piles of limbs outside of operating sites are numerous, objectifying for the observer the almost inconceivable horror of the experience of battle. Almost 30,000 amputations were performed by Federal surgeons—with an overall mortality rate of 26.3 percent.[27] One important factor was the frequency of occurrence of extremity wounds, with the limbs making up the greater part of the exposed target in a stand-up battle line. Another part of the reason for this should now be apparent: it was in the limbs, rather than the head and trunk, that the surgeon had at least an opportunity to improve upon the results of expectant treatment. The decision for amputation resulted from a calculation involving the likelihood of survival without surgery, the likelihood of surviving with the surgery, and the knowledge that delaying surgery past twenty-four to forty-eight hours reduced the likelihood of survival. In addition, unlike civilian hospital-based surgery, the resources available to provide surgery, nursing care, and transportation played a role in the decision. A soldier with an amputation was easier to care for, easier to transport, in much less pain, and more likely to survive a period of rough handling than a soldier with a leg fracture or a resected joint requiring prolonged limb immobilization and close nursing care.[28] The fate of each soldier wounded in an extremity was determined by such grim arithmetic.

The major indication for amputation following extremity wounds was extensive bone or direct joint damage: the "shot fracture."[29] The conoidal "ball" most commonly used had particular ability to fracture and splinter bone. Simple fractures could often be treated without surgery, but extensively pulverized and splintered bone with many small fragments represented a major threat to survival without surgery. The wound would not heal unless the fragments of dead bone were removed, but extensive manipulation and exploration of the wound necessary to accomplish this, in the preantisepsis era, too often led to a poor result, with death from complications of "mortification," including pyemia, hospital gangrene, and secondary hemorrhage. Finally, fractures involving a joint, or fractures associated with a great deal of soft tissue damage, would also prove fatal due to what are now known to be infectious complications if not treated with amputation.

With regard to timing, it was known that when indicated for treatment of the initial injury, delay of the amputation beyond twenty-four hours into the period of established inflammation of the initial wound substantially increased the mortality rate. For example, the mortality rate of thigh amputations rose from 53 percent to 64 percent after forty-eight hours.[30] The results of surgery performed in the period of "wound inflammation" and reaction were so poor that if initially delayed surgery was later deemed necessary, it was delayed if at all possible "until the lesions had become local and analogous to chronic disease."[31]

Arguing against amputation were two main considerations. Location of the proposed amputation closer to the trunk dramatically increased mortality rates. The fatality rate for amputations in the lower leg was 33 percent and in the thigh 53 percent.[32] In the thigh, the closer the amputation was to the hip, the higher the mortality.[33] Amputations for hip fractures carried such an appalling mortality rate (more than 90 percent) that there were rarely performed, despite the untreated mortality of 83 percent.[34] Secondly, the condition of the patient was an important factor. A soldier with symptoms of collapse, including delirium, high fever, and a weak pulse, was much less likely to survive surgery.

Overall, about one half of extremity shot fractures were treated with amputation.[35] Two years after the end of the war, Lord Lister published an account of a compound fracture of the thigh treated by the use of carbolic acid dressings, setting the fracture, and allowing healing to take place. No mortification intruded. The era of antisepsis changed the arithmetic of amputation forever.

Amputations were almost invariably performed under anesthesia, so operator speed was less important than in the preanesthetic days of sixty-second surgery. This allowed more time for careful technique. Anesthesia was light, which allowed for remarkably few anesthesia deaths and quick recovery of consciousness afterwards.

Prior to beginning, anesthesia was induced by having the patient breathe through an inverted cloth cone held over the nose and mouth by an assistant who dripped ether or chloroform onto the cloth. (Chloroform was in fact preferred because it was less combustible than ether, and operations were not infrequently performed by the light of lanterns.) Once the patient was insensible, a tourniquet or simple pressure on the major arteries to the limb interrupted blood flow to the limb to control bleeding during the surgery. One of two surgical techniques was employed. In the circular method, a circular skin incision was made with a scalpel, and a cuff of skin pulled up the limb to allow a similar but deeper incision to be made at a higher level through the muscle down to the bone. The cuff of muscle was then pushed up to allow the bone(s) to be sawed through at a higher level yet. In the flap method, a double-edged knife called a "catling" was pushed through the limb next to the bone, and then used to cut through the muscle and skin in a direction towards the hand or the foot to create a large flap of tissue. The flap was pulled forcibly up, the surgeon dissecting the muscle free from the bone with a knife. After a similar cut and dissection on the other side, the bone was exposed for cutting, flanked by two large flaps of tissue covered with skin. The flap method was faster and easier for a less experienced operator; the circular method was felt to often deliver better results.

In both methods, major arteries were identified by the assistant and fixed with a "tenaculum" hook with a knotted piece of surgical silk looped over it. The knotted loop was moved over the end of the artery and tied tightly. One string was cut, leaving one long piece of the silk attached to the knot. Once the major arteries were tied, the bone was cut through as high as possible and the end filed smooth.

In the circular method, the cuff of muscle was brought down and closed over the smooth end of the bone, providing a thick pad over it. Then the cuff of skin was brought down over the muscle, and the edges fastened together closing the skin over the muscle, after the long ends of the knots tying the arteries were carefully identified, brought through the opening, and fastened to the skin with adhesive plaster. The creation of the thick pad of tissue and skin over the bone end was essential both for healing and the eventual use of an artificial limb.[36] In the flap method, simply joining the edges of the flap over the smoothed end of the bone had the same effect. The patient was removed to have a dressing—held in place by artfully wound bandage rolls—applied. The anesthetic cone having been removed, the patient quickly recovered from the minimal anesthesia depth attained by this technique.[37]

An alternative to amputation that preserved some extremity function, when removal of the injured area was necessary, was resection. In this procedure, the injured part of the bone or an entire joint, was cut away with a hand chain saw that could be passed around the bone without injuring other tissues. The resulting shortened limb was treated as a severe fracture, and could eventually reach a level of some useful function, if only to preserve a useful hand in an upper arm fracture, for instance. There was a great deal of enthusiasm for resection in place of amputation in Confederate publications.[38] The compiled experience of the Union surgeons was not supportive: in the upper extremity results were "disappointing" and in the lower extremity "disastrous," with mortality rates higher than for amputation, especially in the upper extremity.[39] Unfortunately, this surgery required more skilled operators and long periods of absolute immobilization of the limb not possible under battlefield and back-country transport conditions, and took longer to perform. What was appropriate for a civilian surgeon in a large urban hospital might not be appropriate for a military surgeon operating on dozens of patients outside of a tent in a rural field for thirty-six hours without a break. All these factors likely were responsible for the fact that only 14 percent of extremity shot fractures treated surgically were treated with limb-conserving resection.

Once operative treatment was concluded, care of the wound was designed to relieve suffering and promote initiation and completion of the healing process. Despite the elaboration of treatments based on theoretic conceptions of causes of inflammation and "mortification," many of the treatments were empirically based.

Adequate pain control was available to the suffering wounded soldier. The basic medicine kit contained opium-based narcotics in both pill and liquid form, as well as injection syringes used for subcutaneous injections of morphine.[40]

Dressings were intended both to keep the wounds clean and to provide pain relief. Application, inspection, and changing of dressings was largely a task of the surgeon attending the case, or an assistant surgeon. Despite the illustrations of elaborate and beautifully symmetric bandages in minor surgical manuals of the time,[41] the usual dressings were kept simple to facilitate comfort for the patient and convenience for the surgeon. Given the degree of local inflammation almost invariably present in and around the wound, there were theoretical reasons supporting the favored type of dressing: the cold water dressing. The patients luckily found this almost universally to comfort the pain of the wound substantially. The simplest form consisted of folded lint placed over the wound, saturated with cold water, and fastened with a few strips of adhesive plaster, without bandages. When more intense cooling was desired for active inflammation, a bucket with cold water and ice stood near the wound, with water wicked to the lint by stands of cotton to replace evaporation. The dressings were intended to be changed at least daily, and more often when a profuse or foul discharge was present, the dressing-changer keeping the wound clean and checking on the progress of healing. Once inflammation was resolved and only healthy pink tissue present, a simple dressing of lint spread with a fat and wax mixture (simple cerate) was used.[42]

Despite the inflammatory components present, contemporary surgeons recognized that a downward course to prostration and death was the greatest danger to the wounded soldier. Consequently, highly nutritive diets favoring eggs and beef extracts in various forms and tonics such as iron-based tinctures—systemic treatments that would not be used in most obviously "inflammatory" diseases—were the mainstays of supportive treatment.

The ever-present fear for the soldier during the healing process was the specter of "mortification"—increased pus discharge, degeneration, and sloughing of tissue in and around the wound heralding a potentially fatal turn for the worse. A class of preparations known as disinfectants existed, including iodine preparations, bromine, and chlorine solutions. One of these, creosote, in its British but not American formulation, contained significant amounts of carbolic (phenolic) acid, which Lister would soon make the basis of antiseptic surgical and wound care techniques. In fact, reports were available even to Confederate surgeons of European enthusiasm for carbolic acid as a disinfectant.[43]

Unfortunately, the concept of using these compounds as a prophylactic on the skin surface or on a healthy wound had not developed. Disinfectants were defined as substances to render mortifying flesh less foul rather than in the modern sense of substances to prevent mortification. So they were applied to the wounds only once foul mortification was observed, although early—albeit ineffective—preventative thinking was evident in the placing of pans of chlorine solution in the wards to fight contagious complications such as hospital gangrene and erysipelas (discussed below).

Once a soldier survived his wounding and any operative treatment rendered, his ultimate fate depended on successful healing of the wound. This was by no means assured.

The wound might simply fail to heal, with ongoing discharge of pus, failure of the "primary wound fever" to resolve, and a gradually weakening condition leading to decline and death. In retrospect, this was probably most often due to an ongoing infection. Autopsies at the time disclosed on occasion unsuspected retained foreign bodies, bone sequestrum, abcesses, or chronic bone infections. More often, no specific cause was identified. In the weakened state of these patients, the slow decline might be hastened by the development of pneumonia or debilitating diarrhea. Given what we now recognize as the unsterile conditions surrounding the examination, the wound cleaning, and the dressing changes, that this result would sometimes occur is not surprising. Without our knowledge of infectious diseases and their causes, this was viewed not as a complication, but as a natural end of the process that the treatment rendered had not prevented.

Especially feared was the dramatic mortification associated with pyemia, erysipelas, and gangrene, which were recognized as distinct complications that could interrupt an apparently successful recovery.

Pyemia, recognized today as occurring when infection spreads from the wound into the bloodstream, was especially likely to occur following wounds of the long bones and joints. The soldier would be observed to become restless and anxious, pale, and to lose his appetite. This prologue would be succeeded by sudden high fever and shaking chills, progressing within several days to enfeebled circulation and collapse into unresponsiveness and death. On occasion, widespread joint pain, swelling, and redness heralded the spread of infection throughout the body.[44] Once established, the mortality rate was over 97 percent in 2,800 Federal cases. No treatment seemed to help. A rare patient survived when deterioration in the appearance of the wound coincident with the onset of pyemic symptoms led to urgent amputation—but most did not.[45]

The word *gangrene* comes from the Greek root meaning "to eat." The word captures the horror of dissolving flesh that accompanied epidemic hospital gangrene, or "wet" gangrene. The disease could intervene at any point in the recovery of a wound in a hospitalized patient. The initial symptom was the development of a burning or pricking pain in the wound, which quickly became severe and intense. At this point an ash-colored layer adherent to the surface of the wound would appear, accompanied by a watery discharge. The edges of the wound became discolored in a livid or violet hue, and became hard and firm. From this point on, the wound would enlarge almost visibly from hour to hour, with the rapidly dissolving flesh undermining the expanding margin of the wound.

Treatment of hospital gangrene, ineffective at the start of the war, by the end was effective in individual cure as well as preventing spread to new cases. Spread was prevented by rapid isolation of cases, and taking care not to exchange dressing materials among patients, or even among wounds on one patient. Treatment involved using chemical agents to literally burn the infected wound area once each day, removing the scarred, infected tissue between applications, and repeating until only healthy, pink scar tissue was present. The most successful chemical application was liquid bromine, which had to be applied under anesthesia because of the initial pain.[46] Although clearly to modern eyes an infectious disease, the precise bacterial cause of nineteenth-century hospital gangrene remains unknown.

Erysipelas, now known to be an infection of the skin and superficial tissues by a virulent and highly contagious streptococcal bacterium, was known to occur both spontaneously and as a complication of wounds. The usual picture was the abrupt appearance of a bright red discoloration of the skin surrounding the wound and a high fever. The involved skin area spread rapidly to adjacent parts. Although cases were seen where the entire body became involved, the disease tended to be limited in extent. Although over 40 percent of soldiers who developed erysipelas died, it was felt to be the direct cause of death in only 15 percent of the cases, with the other deaths attributed to hemorrhage, pyemia, and other diseases. The highly contagious nature of the disease was known, and, as with hospital gangrene, outbreaks could be controlled with isolation and rigorous care forbidding the use of cleaning and dressing materials on more than one patient.[47]

Treatment of this violent inflammatory disease suggested by the contemporary medical systems of thought would be various depleting therapies promoting purgation and sweating meant to decrease the "excitement of the vascular system." However, experience showed that patients either recovered rapidly on their own or became prostrate, so support and stimulation with a nutritive diet and tonics were substituted. Tincture of iron was felt to have some specific effect, and the skin lesions were treated locally with tincture of iodine or silver nitrate applied to the lesions, or in a band of sound skin surrounding the involved area in the manner of a firebreak. Sometimes this seemed to work.[48]

Perhaps the most feared common complication of a shot wound was secondary hemorrhage. This was any hemorrhage in the area of the wound occurring once the free discharge of pus from the wound had become established—usually within several days of the wound or surgery. It was recognized that the usual cause, verified at autopsy, was "sloughing" of the artery wall. Unrecognized, of course, was the infectious cause of the arterial damage.[49]

The mortality of treating secondary hemorrhage was high. Emergency ligation of the supplying artery was necessary, often high above the wound to control the bleeding. If this control, once obtained, so compromised the blood flow to an extremity that it became non-viable (dry gangrene), urgent amputation was necessary. Sometimes, in fact, emergency amputation—or reamputation—was necessary to control the bleeding.

This complication was especially significant following an amputation. The arteries were divided and tied off with silk at the time of the surgery, as earlier mentioned. The ligatures, however, constituted foreign bodies in the wound, and complete healing would not occur until they were removed. Sterile ligatures, antiseptic operating technique, and absorbable suture material would eliminate this problem in the future. At this time, however, all the surgeon could do was gently tug on the strings brought out through the wound beginning ten days or so after the surgery, and do so daily until each came loose and could be removed. The sloughing process that always occurred, ideally, would loosen the tied end of the artery without weakening or destroying the clot and scar that plugged the artery, allowing the ligature to be removed and final healing of the wound to occur. The unknown enemy of bacterial infection was, therefore, a tenuous ally in this process that could turn fatal with each tug of the string. The anxiety and circumscribed drama attending this task—especially on the part of the patient—can easily be imagined.

At some point in this journey from wound to healing, the soldier would either be returned to duty or be transferred to one of the fixed general hospitals, well behind the lines and generally placed in the large cities of the North and the South. In these often newly built pavilion-style hospitals constructed hurriedly of rough wood, the struggle to survive would continue.

Fewer than two-thirds of the wounded ever reported back to their regiments to face the odds again. For the other survivors taken off the field, the war was over, one way or another. About 11 percent of those hit got off the field alive but eventually succumbed to their wounds or complications—or 14 percent of the wounded that entered the army medical system.[50] A similar proportion of the wounded was sent home from the general hospital, unable to return to duty due to disability.[51]

The results, despite our modern judgments, were not primitive; the success of the Union (and Confederate) army medical system represented clear improvement over those of the recent European wars, and equal or superior to the European wars to come in the following ten years.[52] If the participants did not anticipate years of medical progress to come, their achievement under difficult circumstances is not diminished.

Those disabilities, reminders to the soldiers and their society of the violence of war, would generally accompany them for the remainder of lives often shortened by their effects. Disability was a polite word for:

- Mangled faces, imperfectly disguised by mustaches or beards
- Useless or impaired limbs
- Literally tens of thousands of missing body parts (One-fifth of the 1866 State of Mississippi income was spent on prosthetic limbs for former soldiers.)[53]
- Chronic pain from wounds, imperfectly healed, or agonizing, boring, burning pain from nerves mangled by lead
- Wounds discharging fluid and pus for years, occasionally spitting up small pieces of bone or barely recognizable objects such as buttons or nails

These were not the results anticipated when these men went to war—glorious battles, a quick and valiant death for the unlucky few. Instead, they paid a high and distinctly inglorious price for upholding their principles at the cost of their bodies.

The wounded and the surgeons embarked on this unexpected journey together. But for the surgeon, the drama was repeated, patient after patient, day after day, month after month, until it must have seemed as if the wreckage of battle would consume the nation. An army surgeon didn't hold romantic illusions of war and battle. One wrote: "[The surgeon] . . . understands what the soldier's life finally brings to many, the death wound, the burning fever, the wasted body, and the broken constitution. He knows what battle means—the shattered limbs, the moan of pain, the life long cripple."[54]

Notes

1. Oliver Wendell Holmes, Jr. *The Essential Holmes: Selections from the Letters, Speeches, Judicial Opinions, and Other Writings of Oliver Wendell Holmes, Jr.,* edited by Richard A. Posner (Chicago: University of Chicago Press, 1992), 87.

2. William F. Fox, *Regimental Losses in the American Civil War* (Albany, N.Y.: Brandow Printing Co., 1889; repr., Dayton, Ohio: Morningside, 1985), 47. All citations are to the 1985 edition.

3. John W. DeForest, *Miss Ravenel's Conversion from Secession to Loyalty* (New York: Harper & Brothers, 1867; repr., New York: Penguin Books, 2000), 408. Citations are to the 2000 edition.

4. Ibid., 258.

5. Fox, *Regimental Losses,* 47. This and most of the statistics in this article are taken from Union records, due to the incomplete and uncertain nature of comparable Confederate material.

6. *The Medical and Surgical History of the War of the Rebellion, (1861–65)* (hereafter, "MSH"), 6 vols, vol. 2 (surgical volume), part 3 (Washington: Government Printing Office, 1870–1888); as *The Medical and Surgical History of the Civil War,* 12 vols. (Wilmington, N.C.: Broadfoot Publishing Company, 1990–1991), 704–705. All citations are to the 1870–1888 edition.

7. Ibid., 709–710.

8. Ibid, 761.

9. Ibid., 696.

10. Ibid., 759.

11. Ibid., 760.

12. Mary C. Gillett, *The Army Medical Department 1818–1865,* edited by David F. Trask, *Army Historical Series* (Washington, D.C.: Center of Military History, United States Army, 1987), 153.

13. George Worthington Adams, *Doctors in Blue* (Baton Rouge and London: Louisiana State University Press, 1996), 47.

14. Ibid., 46.

15. Jonathan Letterman and Lt. Col. Bennett A. Clements, *Medical Recollections of the Army of the Potomac and Memoir of Jonathan Letterman, M. D.* (New York: Appleton, 1866; repr., Knoxville, Tenn.: Bohemian Brigade Publishers, 1994), 162. Citation is to the 1994 edition.

16. MSH, vol. 2, part 3, 887.

17. J. Julian Chisolm, *A Manual of Military Surgery* (Columbia, S.C.: Evans and Cogswell, 1864; repr., Dayton, Ohio: Morningside, 1992), 172. All citations are to the 1992 edition.

18. Ibid., 53–58, 696.

19. Ibid., 688–690.

20. Gordon C. Rhea, *The Battle of the Wilderness: May 5–6, 1864* (Baton Rouge and London: Louisiana State University Press, 1994), 365.

21. Ibid., 441–442.

22. Emory M. Thomas, *Bold Dragoon: The Life of J. E. B. Stuart* (New York: Harper and Row, 1986), 292–295, 300.

23. Alice Rains Trulock, *In the Hands of Providence: Joshua L. Chamberlain & the American Civil War* (Chapel Hill and London: University of North Carolina Press, 1992), 2–5, 466.

24. MSH, vol. 2, part 1, 719.

25. Ibid., vol. 2, part 3, 761–766.

26. Ibid., vol. 2, part 1, 719.

27. Ibid., vol. 2, part 3, 877.

28. Chisolm, *Manual*, 407–408.

29. Ibid., 409–410.

30. MSH, vol. 2, part 3, 213.

31. Ibid., 879.

32. Ibid., 213, 461.

33. Chisolm, *Manual*, 481–483, 66–68.

34. MSH, vol. 2, part 3, 65.

35. Ibid., 873.

36. Chisolm, *Manual*, 481–483, 66–68.

37. Steven Smith, *Hand-Book of Surgical Operations* (New York: Bailliere, 1862; repr., San Francisco: Norman Publishing, 1990), 92–95, 42–45, 26–32. Citations are to the 1990 edition.

38. John Stainback Wilson, "Resection of Upper Half of Humerus," *Confederate States Medical and Surgical Journal*, vol. 1, no. 4 (1864), 3–5.

39. MSH, vol. 2, part 3, 873.

40. George Winston Smith, *Medicines for the Union Army* (Madison, Wisc.: American Institute of the History of the Pharmacy, 1962), 114–115.

41. John Hooker Packard, *A Manual of Minor Surgery* (Philadelphia: J. B. Lippincott & Co., 1863; repr., San Francisco: Norman Publishing, 1990), 104, 106, 111. Citations are to the 1990 edition.

42. Joseph Janvier Woodward, *The Hospital Steward's Manual* (Philadelphia: J. B. Lippincott & Co., 1862; repr., San Francisco: Norman Publishing, 1991), 300–303. Citation is to the 1991 edition.

43. "Interesting Analyses of Articles in Foreign Journals," *Confederate States Medical and Surgical Journal*, vol. 1, no. 11 (1864):191.

44. Chisolm, *Manual,* 248–249.

45. MSH, vol. 2, part 3, 858.

46. Ibid., 823–851.

47. Ibid., 851–857.

48. MSH, vol. 1, part 3, 763–764.

49. MSH, vol. 2, part 3, 809.

50. Fox, *Regimental Losses,* 47.

51. MSH, vol. 1, part 3, 646–658, 716–718.

52. Louis C. Duncan, *The Medical Department of the United States Army in the Civil War* (Gaithersburg, Md: Old Soldier Books, 1985), 407.

53. Thomas L Connelly and Barbara L Bellow, *God and General Longstreet: The Lost Cause and the Southern Mind* (Baton Rouge: Louisiana State University Press, 1982), 8.

54. Jacob Ebersole, "Letters of Surgeon Ebersole, 19th Indiana," (Frederick, Md: Collection of the Museum of Civil War Medicine).

4

"The Awful Shock and Rage of Battle"

Rethinking the Meaning and Consequences of Combat in the Civil War

Eric T. Dean

The Prussian officer and philosopher of war, Carl von Clausewitz, is well known for his dictum that war is merely the continuation of politics by other means. In the words of Clausewitz himself: "War is an instrument of policy. It must necessarily bear the character of policy and measure by its standards. The conduct of war, in its great outlines, is therefore policy itself, which takes up the sword in place of the pen."[1]

What is so shocking and controversial about these words is that Clausewitz was attempting to argue, in essence, that war is not a horrible anomaly, which results in wide-scale death and destruction, and leaves thousands of widows, orphans, and mutilated or crippled men in its wake, but, rather, that war is somehow logical, rational, and unavoidable, a mere continuation of political controversies and struggles. In the Clausewitzian view, the decision to wage war can be seen not as a breakdown of decency and a departure from sanity, but as a legitimate or even reasonable way of asserting or vindicating one's interests. The rhetoric of Clausewitz can be stirring and frightening: "War is thus an act of force to compel our enemy to do our will . . . there is no logical limit to the application of that force. Each side, therefore, compels its opponent to follow suit; a reciprocal action is started which must lead . . . to extremes . . . Every engagement is a bloody and destructive test of physical and moral strength. Whoever has the greater sum of both left at the end is the victor . . . Direct annihilation of the enemy's forces must always be the *dominant consideration.*"[2] Writing in the early nineteenth century, Clausewitz seemed to almost revel in describing Napoleon as the "God of War himself," with "unlimited driving power." He referred to the French army's "pulverizing course through Europe," and noted: "Woe to the government, which, relying on halfhearted politics and a shackled military policy, meets a foe who, like the untamed elements, knows no law other than his own power!"[3]

Leaving aside for the moment some of the complexities of his thought, what I would like to suggest is that Clausewitz's basic outlook on war as logical and the mere continuation of politics by other means has perhaps come to pervade contemporary thinking on the American Civil War, in a way that few have appreciated, and in a way that has tended to rationalize it, thereby overlooking and minimizing the stark brutality of battle, and the effect this has on soldiers and their families. In a nutshell, one sees Clausewitzian logic brought to bear—by means of the New Social History—and with this synthesis becoming the predominant paradigm in Civil War scholarship. Where Clausewitz saw war as the continuation of *politics* by other means, Civil War historians have, in the past several decades, tended to view the Civil War as a continuation of *society* by other means, or, if you will, the continuation of social discourse by other means.

At the outset the New Social History must be placed into some sort of meaningful context. Historians in the nineteenth century, such as the British historian Thomas Babington Macaulay, focused on political and military history; in a sense, these historians viewed things "from the top down," and interpreted history as being defined by important events and dates, and saw these historical events as being driven by the efforts of great men and women such as kings and queens, or by statesmen and generals, such as Bismarck and Napoleon. By attempting to explain such upheavals as the French Revolution or the rise of nation states in Europe, these historians regarded the state as the chief agent of historical change, and, indeed, the work of these historians was often supported or sponsored by the state, unlike today's modern professional historian, who generally finds his or her home in a university setting.[4]

In the late nineteenth and early twentieth centuries, however, historians such as J. R. Green and J. H. Robinson became dissatisfied with the predominantly political, constitutional, and military emphasis in history, and introduced what became known at that time as the "new history," something that has evolved, especially since the end of World War II, into a fascination with social history; this type of history focuses on the study of the common man, which generally seems to include all common people such as peasants and laborers, but, also those traditionally thought of as "deviants" or "outsiders," such as strikers, rebels, criminals, and madmen. To social historians, laws and institutions are social constructions, invented by elites and intended mainly to control or oppress the masses; one sees a discourse or dialogue emerge in the New Social History in which the common man, or especially the common woman, engages in a struggle to determine his or her own fate: hence, the rhetoric of struggle and political confrontation in the small matters of daily life. If you peruse current book reviews in a journal such as the *Journal of American History*, you will quickly see that social and cultural history have become all the rage.[5]

In the realm of Civil War studies, the New Social History centers around the study of community, gender, and the common man, with the idea that the common man or woman was not a passive object, but rather that this person was empowered and autonomous, struggled against and contested oppression and unreasonable authority, and, on a certain level, was almost heroic. These principles are seen at work, perhaps most dramatically, in African-American studies. Prior to the 1960s, historians relegated the approximately four million slaves in the Old South in addition to free blacks in the North to a marginal status, and regarded these people as pawns or victims, who did not play a direct or important role in the American Civil War.

However, in the past several decades, scholars such as Ira Berlin, Barbara Fields, and Joseph Glatthaar have rewritten history to grant African-Americans a pivotal role in the war. Three central components form the basis for this new interpretation of the role of African-Americans in the Civil War. First, the revised argument on emancipation is that the slaves were not freed by an external force in the North, but that they liberated themselves by sabotaging southern agriculture, running away when the opportunity permitted, or enlisting to fight in the Union Army—hence, the theme of self-emancipation and autonomy, with the underlying assertion that these African-Americans contested the power of white elites. Second, historians such as Ira Berlin no longer regard politicians such as Senator Charles Sumner of Massachusetts or President Abraham Lincoln as visionary trailblazers who led the way to emancipation and civil rights through the crafting of laws and edicts such as the Civil Rights Acts or the Emancipation Proclamation; rather Sumner, Lincoln, and their colleagues are viewed as being *forced* to pass civil rights legislation in response to the new reality of African-American freedom. In this view, northern politicians were simply bringing the law into a state of congruence with social reality. And third, the approximately 180,000 African-Americans who fought for the Union are seen as playing a pivotal role in victory; they entered the war at a time when northern manpower and resolve were waning, and saw the cause through to victory. In the words of

Ira Berlin: "Through their behavior, slaves compelled Federal authorities to adapt their policies to match the increasing magnitude of the war. . . [The slaves] watched and waited, alert for ways to turn the military conflict to their own advantage, stubbornly refusing to leave its outcome to the two belligerents. . . In throwing off habitual restraints, freedpeople redesigned their lives in ways that spoke eloquently of their hidden life in bondage." The key words here are "compelled" and "redesigned." The key concept is the idea of the common man determining historical events, with the theme of this common person (usually defined as women and minorities) not as a victim, but as autonomous agent and heroic master of her own fate, engaging in perpetual conflict with social and political elites, usually defined as white males.[6]

The outlook and approach of the New Social History as reflected in African-American scholarship have spilled over into the study of warfare itself, and, particularly, into the study of the Civil War soldier. This approach to the study of the Civil War contains grave shortcomings or dangers, and I would like to discuss three representative and influential books that sharply demonstrate these principles at work. First of all, Reid Mitchell's *The Vacant Chair*, written in 1993, explicitly attempts to merge gender and family studies with the traditional concerns of Civil War studies; his object is to show that the experience of the Civil War soldier was shaped by images of home and the family, and he thereby attempts to grant a greater role to women, who have often been viewed as bystanders in the midst of warfare. His analysis contains two key points, the first of which is that gender was a dynamic force, and, second, that the masculine in and of itself was inadequate and that the feminine was critical if not determinative. Mitchell makes this argument by investigating the nature of life in the service, and concludes that there could be no such thing as a purely masculine world: "The lack of feminine presence meant that the masculine world of the army was incomplete." He hypothesizes that in order to function in the masculine world of army life, men found that they themselves had to embody some virtues thought to be feminine; to a certain extent, they had to act like women, by literally doing women's traditional work such as cooking, washing, and mending clothes, and in other respects, by taking on the role of nurturing: "The preeminent masculine pursuit of war forced men to develop some of the feminine virtues within themselves. 'Men without women' had to learn at least some of the virtues they had relegated to the domestic sphere."[7]

In the second facet of his work, Mitchell goes beyond arguing the mere inadequacy of the masculine, by positing a dynamic, formative influence of the feminine through the agency of common women at home, whose influence operated at a distance. Mitchell does this by revisiting the theme of the Civil War as a kind of "coming of age ritual" for the men who fought the war; such a ritual has traditionally been regarded as in the purview of males, who bonded together and fought in groups marked by fierce loyalty to one another. However, Mitchell reinterprets this process by exploring the definitions of masculinity that shaped this passage, and finds that the young Civil War soldier was wrestling with notions of manhood such as "manly restraint," "self-discipline," and "civilized morality," which were largely defined and set forth for him by the community and family at home. Mitchell sees a process whereby the Civil War soldier achieved manhood as a kind of domestication in which the feminine, domestic sphere served as a ground for the masculine, public world.[8]

Thus, Professor Mitchell advances a thesis of "connectedness," by suggesting that a close and continuing relationship existed between the Civil War soldier, and his mother, sisters, and community, a relationship that shaped his behavior and development during the war. What, you may ask, is deficient about this study of and approach to the Civil War soldier? Before evaluating Reid Mitchell's thesis in *The Vacant Chair*, we should return briefly to Clausewitz's magnum opus, *On War:* "The aim of warfare is to overcome or disarm the enemy; war, however, is not the action of a living force upon a lifeless mass, but always the collision of two living forces . . . Every engagement is a bloody and destructive test of physical and moral strength."[9]

What Mitchell and others who extend community and gender studies to the Civil War miss is that warfare, for the participants "on the ground," is a shocking, horrific, and wrenching experience—which may lead to a kind of alienation from all that is normal, civil, and decent. When sent into combat, the Civil War soldier encountered a surreal landscape of smoke, noise, and confusion. Accounts of battles routinely refer to the "blinding smoke," the "awful din," the "terrific and deafening" roar of gunfire, and the "shrieking" of artillery shells. Memoirs written even decades after the war recall the "wild havoc" and the "awful shock and rage of battle."[10] As one participant at the Battle of Gettysburg wrote: "The troops taking part were sweaty, blackened by the gunpowder, and they looked more like animals than human beings. The animal-like eagerness for blood, the need for revenge, painted a terrible picture of white faces and bloodshot eyes. This portrait of battle was a portrait of hell." As another Civil War soldier wrote: "Why will *man*—created in the image of God, act like a fiend incarnate? . . . Why was man created, and allowed to perpetrate such wickedness?"[11]

Noting that battle could shock men with incredible force, Earl Hess has characterized the experience that Civil War combat soldiers endured as "crossing over": "reality proved to be a grisly shock. . . . They had to cross over the gulf that separated naive imagination from brutal reality in order to understand combat . . . A gigantic gulf existed between those men who had been in combat and those who had not . . . [Men exposed to combat] crossed over the gulf of experience, leaving behind relatives and friends who could not know what had happened to them. . . Once they got through the searing experience of combat, survivors found themselves set apart from other people, even those closest to them."[12]

Not surprisingly, a common theme that emerges in the letters or memoirs of Civil War soldiers is that of alienation from those at home, and the conviction that their experiences could not be understood by those who had not personally been in combat.[13] In describing a battle, one Union veteran wrote: "The genius of Dante could but faintly portray the horrors of that hell of fire and sulphurous smoke—the crash and roar of artillery and musketry . . . the agonizing shrieks of those wounded from the bayonet thrust . . . or crushed by fragments of exploding shell, sinking to earth a mass of quivering flesh and blood in the agony of horrible death. The half can never be told—language is all too tame to convey the horror and the meaning of it all."[14]

This belief that civilians simply could not understand and that language could not do justice to the reality they had witnessed was particularly pronounced when Civil War soldiers tried to relate what they saw on the battlefield after the firing had ceased. Such attempts to describe the carnage of a battlefield strewn with dead bodies and assorted wreckage seemed inevitably to end with phrases such as "Pen cannot properly describe this valley of death, it was too horrible"; "the horrors of a battle field cannot be described, they must be seen"; "The most shocking sight I ever saw"; "ghastly"; "O what a sight, it almost makes me shudder to think of it."[15]

Given that veterans "crossed over" into another world, and thereafter felt alienated from civilians at home, the more interesting question that social historians might address is not how these men related to those at home during the war, but how they readjusted to civilian life *after* the war. Noting that the veterans' movements in the North and South did not become active until the 1880s, some historians have suggested that Civil War veterans came home after the war, returned to civilian pursuits, and didn't really think much about or dwell on their wartime experience for another twenty years. I seriously doubt that this was the case. One should ponder for a moment the Winslow Homer painting entitled, "The Veteran in a New Field."[16] This work depicts a Union veteran working in a field at harvest time; he has set aside his army jacket and is in his shirtsleeves, and only his back is visible as he is stooped over, swinging his scythe. He is alone in the field, alone with his work and alone with his thoughts. What were these thoughts? One of the challenges for Civil War historians is to discover sources which will more fully reveal the story of the returned veteran in the years directly after the war.

The second book I would like to discuss is Earl Hess's *The Union Soldier in Battle*. Written in 1997, this book demonstrates great insight in many respects, but a critical aspect of the work resonates, intentionally or unintentionally, with the flawed tendency of the New Social History to portray the common man as autonomous, empowered, and almost heroic in his struggle against the social and political elites, or external circumstances. Hess begins his study by describing with great power the shock and chaos of battle, and the way in which this tended to produce an identity for northern soldiers, separate and foreign from the outlook of those on the home front. However, his narrative takes a strangely optimistic turn when it addresses the issue of how the Union soldier coped with these stresses.[17] Hess argues that the Union soldier resisted becoming a passive victim of war, and, through considerable efforts of self-control, became instead a *victor* over the horrors of combat. This involved, first of all, gaining some degree of control over his emotional response to combat by using models and experiences from his civilian life to understand the experience of battle. Second, the soldier strived to create a coherent vision of battle by shaping his perception of the event, and in the process defined courage and his own role in the war. Last of all, most Federal soldiers did not allow frustration, bitterness, or callousness to permanently alter their character or their faith in the war effort. In the final analysis, Hess presents this struggle as a saga of courage, the story of men fighting to overcome cowardice and emotional failure:

Only by effectively dealing with the fear, excitement, horror, and exaltation known by the warrior could they *become real soldiers* . . . [as the soldier entered battle] he was *forced to find the limits of bravery and of cowardice within himself,* to delineate his own personal field of battle . . . Most of the men who flocked to the regimental recruiting stations came through this personal reckoning intact . . . Most, but not all, *endured the test.* They defined themselves as soldiers capable of dealing with the worst that combat had to offer . . . Defining courage was a task that all Union soldiers had to undertake . . . Their success was the key to saving the Union . . . Inevitably, some proportion of the Northern army *failed the emotional challenge of battle,* but the exact percentage is impossible to determine.[18] [emphasis added]

What concerns me about this interpretation of brave men becoming real soldiers and saving the Union, and cowards failing the emotional challenge of battle, is that it runs contrary to our discoveries in the twentieth century that men in battle can only take so much before they can take no more, and become psychiatric casualties. At the beginning of the Second World War, the U.S. Army adopted the theory that if so-called "defectives" were adequately screened out at induction, then those men who had passed muster and remained in the ranks would be resistant to or immune from psychological breakdown. Although Army induction centers rejected over 1 million men due to supposed psychiatric reasons in World War II, the policy clearly failed as hundreds of thousands of men in the field still fell victim to battle fatigue. The Army eventually recognized that "resilient" or "immune" soldiers were nonexistent and that "every man has his breaking point"; it came to understand that if *any* man were exposed to enough fear and death, he could be reduced to tears and uncontrollable shaking, as was portrayed in the deeply moving motion picture, *Saving Private Ryan.* Accounts of the breakdown of good soldiers in World War II can be striking: again and again, alarmed psychiatrists described in their reports soldiers who had been exposed to too much combat, and were now completely dysfunctional. Some of these men were dazed and confused, or hopelessly apathetic and depressed; others were tense and anxious and could not settle down or concentrate, and could not sleep at night; or if they did manage to drift off, they experienced horrifying nightmares in which they were once again in battle, watching friends being killed and dismembered before their very eyes. As one psychiatrist wrote: "Some patients return over and over again to one short traumatic scene, living it through repeatedly as if, like a needle traveling around a cracked record, they could not get past this point."[19]

While Hess is correct that war is a complex phenomenon and that it would be wrong to view soldiers in the Civil War era as mere victims, my own study of Civil War veterans, based on a careful review of insane asylum and pension records, reveals many men who were tossed and shaken by forces well beyond their control, and were unable, despite their best efforts, to deal with the consequences of combat. Two specific cases demonstrate war's dreadful repercussions. Dixon Irwin fought with the 13th Indiana Infantry at Chester Station, Virginia, where his unit was bombarded by Confederate artillery. Irwin was knocked off his feet during this barrage, and became disoriented on the battlefield. In a subsequent engagement, a fellow soldier noticed that in the midst of battle Irwin had become overheated and was foaming at the mouth; this comrade gave him water and helped him to the rear, but commented later that Irwin seemed to remain excitable, and acted curious and a little wild thereafter. Although he had been a good and reliable soldier, the impact of continuing exposure to battle wore on Irwin: he became morbidly melancholy, and would go off and sit by himself. Several of his bunkmates had been killed in action, and Irwin seemed to think that he had brought this fate to them; convinced that he was bad luck, he refused to have another bunkmate. He confided to a close friend that he was afraid he was losing his mind.

After the war, Irwin returned to Indiana, married, and had two daughters, but he could not shake his wartime memories. At night, he became timid, and would bar the doors. He said he could hear whisperings around the house, and he became wild and excitable and imagined some one was following him and would kill him. He frequently talked of his company being badly cut up and how they had been shelled during a retreat; he would rant and ramble about these memories of artillery bombardment. A local doctor who examined Irwin subsequently testified: "I remember that his eye had a peculiar appearance as a man who is frightened, and he spoke of the damn big guns. Whenever he spoke of the cause of his trouble he said it was the constant roar of the guns in the service . . . He was wild and very excitable and imagined persons were after him and upon the firing of a gun he was frantic." Irwin's anger and frustration spiraled out of control. In the field, he almost beat his horses to death on one occasion, and at home, he became abusive towards his wife, and threatened to kill her; she finally divorced him in 1874. His family tried to take care of him, but was ultimately unable to deal with his violent spells. Hence, Dixon Irwin spent the rest of his life confined in jails, poorhouses, and insane asylums.[20]

The second case is that of William H. Guile, of the 63rd Indiana Infantry. At the Battle of Atlanta, in July of 1864, Guile's unit was shelled; at least one man was killed, and Guile had to be helped to the rear. When asked about the incident later, he stated that his mind was a blank and that he couldn't remember anything; his comrades in the service noted that thereafter Guile seemed "addled." After his return home to Indiana at the end of the war, neighbors and family observed that Guile was distracted and peculiar, that he was obsessed with his army service, and that he acted alternately violent or gloomy, despondent, and withdrawn. At times, Guile would ramble in his speech and fly from one subject to another, and suddenly break out laughing for no reason; at other times, he would wander around the countryside or shake uncontrollably. Everyone could see that he was not of sound mind. As a neighbor testified to pension bureau officials: "He looks very wild out of his eyes at times, . . . He can not talk with any body two minutes without talking about the war." Most disturbing were his violent spells; he would carry a knife and revolver and say that he wanted to kill someone. At other times, he would get up in the middle of the night and wander around the house with a hatchet.[21]

Nightmares and memories of the violence and brutality of war plagued and haunted many other Union veterans. In fact, one of the most striking symptoms in these men was the fear of being killed. When Elijah Boswell was committed to the insane asylum in Indianapolis in 1871, the commitment ledger noted that he sobbed and cried, and imagined that some one was going to kill him. His brother reported that Boswell thought that the rebels were after him, and that he was in dreadful danger, and, hence, tried to run away. The inquest papers for Michael Cassidy, who had suffered a gunshot wound at the Battle of North Anna River, revealed that he was restless and sleepless, that his eyes had a wild, unnatural expression, and that he was afraid all of the time, and was convinced that someone was trying to kill him. Many other asylum entries reveal the same range of symptoms and behaviors. Here is a sample from a number of case histories: "imagined that some one was trying to kill him"; "Delusion that some one is after him trying to kill him, and has tried to commit suicide by cutting his throat to escape"; "Fearful that persons will kill him"; "Has fortifyed [sic] his house with himself and a Navy revolver . . . delusion that he is holding a fort in state of siege"; "He thinks there is some person after him to kill him"; "In dread of being killed, . . . he will not sit with his back turned to an open door or window for fear that some one will make an attempt on his life"; "says he is afraid of being captured and murdered . . . keeps the doors of the house barred, or fastened at night"; "keeps his fire arms in readiness for self defense"; "he put his axe under his bed at night to defend himself"; "[he] said for me for God's Sake to help him [get] away . . . [because] he thought someone was going to shoot him"; "calls for his gun & . . . talks as if the Rebels were threatening an attack."[22]

For these men and many others, the short and simple was that the situation was completely out of control; there was nothing they, their families, or their physicians could do to deal with the consequences of horrendous memories of death and destruction from the war. Contrary to the theory that Civil War veterans returned home after the war to enthusiastic parades or adulation, readjusted well to family and work, and didn't really think much about or dwell on their wartime experience for another twenty years, I would suggest that many of these men were in profound mental and emotional turmoil. We need more excellent studies of the Civil War soldier along the lines of Earl Hess's *The Union Soldier in Battle*. However, we should augment Hess's analysis of courage and the devices that Union soldiers used to adapt to a surreal and violent world of combat, with additional study of the long-term effects that war had on Union and Confederate soldiers and veterans. Although Civil War sources have been mined, sifted, and scrutinized in detail for the past hundred years, the insane asylum records in most state archives have not yet been systematically studied, and, in fact, have been barely touched in reference to the Civil War soldier and veteran. Much of the story remains to be told.

The third work to be examined is *Divided Houses: Gender and the Civil War*, a collection of essays edited by Catherine Clinton and Nina Silber, first published in 1992.[23] Whereas Reid Mitchell in *The Vacant Chair* investigated the *connections* between women at home and men at the front, the essays in *Divided Houses* seize upon the theme of conflict between the sexes, and investigate the ways in which women in the Civil War era were exploited, and, in turn, contested the power of white male elites. These themes emerge in two major ways: first, in an essay by Kristie Ross on

Union nurses, the emphasis is on the way in which war offered an opportunity to women to escape the domestic sphere, and achieve a kind of self-realization and solidarity with other women. Ross sees these ambitions and opportunities as enthusiastically welcomed by the Union nurses, but as largely thwarted by a white male patriarchy, which resisted the notion of independent and self-reliant women.[24] Second, Drew Gilpin Faust's essay on women in the Confederacy argues that southern women were duped into doing men's work and supporting the war effort by a cleverly crafted myth of sacrifice, created and honed by articulate male southerners. Faust sees this as a form of false consciousness and a "hegemonic ideology," which managed to keep women in line, until things eventually fell apart, and women abandoned the war effort, with dire consequences for the Confederacy.[25]

While new explorations into the meaning of gender in the Civil War era are welcome, this emphasis on conflict, separate identity, and notions of exploitation, hegemony, and a cynical and manipulative white male patriarchy tends to obscure or overlook the way in which the phenomenon of war pulled communities and families *together* rather than apart. In a much underrated book of great insight and perception entitled *The Veteran Comes Back*, the sociologist Willard Waller wrote in 1944 of the solidarity that soldiers and society at large feel in the midst of war: "People discover their love for their country, and lose their puny egos in devotion. Hatred for the enemy binds men together with strands of steel. In the release with full social approval of the usually tabooed emotion of hostility, people find one more thing in common, one more joy of fellowship."[26]

In my investigations of Civil War soldiers, the women who appear in the story are almost invariably deeply concerned about the fate of their sons, brothers, and husbands. The idea that they would have viewed the war as an opportunity for themselves, or that they felt or were indeed in any sense duped or used by shadowy elites of white males seems a peculiar notion. For instance, the commitment records at the Illinois State Hospital for the Insane at Jacksonville reveal a story of distraught and anxious women, deeply worried about the fate of their menfolk. In 1862, Lydia Davis was committed to the asylum, and deemed to be insane; the commitment ledger stated: "Caused by anxiety about husband who is in the army. Has talked of suicide. Is quite violent." For Millicent C. Loar, the ledger reads: "Insane five days. No cause known. Has been anxious about her brothers, two of whom are in the army." In some cases, women had to deal with the death of their husbands, sons, or brothers. Maria Walsh, a twenty-three-year-old widow from Peoria County, was committed to the asylum in 1864, and the examining doctor found that she had been insane for three months; the ledger reads: "Her husband entered the Army about one and a half years [ago] and was killed at Mission Ridge. She has not been right since he enlisted but was not violent until after he was killed." In other Union states and in Confederate states as well, one notices dozens of women committed to asylums both during and after the Civil War when they could no longer handle the grief over having lost husbands, sons, and brothers in battle.[27]

Another important body of evidence concerning gender is that which relates to the way in which women and families cared for Union veterans in the years after the Civil War. While some disturbed veterans became violent and abused their wives, often leading to divorce, the story in many other cases is one of devotion and care. James B. Farr of the 33rd Indiana Infantry had been shot through the neck by a Confederate sniper at the Battle of Kenesaw Mountain. While no account of Farr's reaction to being wounded remains, a Confederate who suffered the same fate of being shot by a sniper related the shock of the event:

> At this time I was shot by a sharpshooter who had crawled within a short distance of the works. I was sitting down . . . when all at once I felt a terrible shock and with a sinking consciousness of dying, became insensible. In an instant I recovered my senses, and found myself with my head fallen forward on my breast, and without power to move a muscle. I could hear the blood from my wound pattering on the ground, and thinking that I was dying . . . I felt *so weak*, so powerless, that I did not know whether I was dead or not. The noise of the battle, seemed miles away, and my thoughts were all pent up in my own breast. My system was paralyzed but my mind was terribly active. My head was full of a buzzing din, and the sound of that blood falling on the ground seemed louder than a cataract. I finally recovered the use of my tongue and . . . told the boys that it was no use to do anything for me, that I was a dead man.[28]

In the case of James Farr, the physical and mental shock of being cut down by sniper fire was doubtless horrific. The Confederate sharpshooter's bullet entered the left side of his body, tore across his neck, and exited near his right ear. Farr was immediately carried to the hospital, where he remained until he was eventually discharged from the army and sent home. His neck muscles were permanently injured, so that he could not bring his head forward in a normal way, but other than this inconvenience, Farr did not seem to suffer any other physical disability; however, the psychological impact of this event was catastrophic. Any physical exertion seemed to cause dizziness and a kind of panic, and neighbors and doctors described Farr as nervous, depressed, and forgetful. As one witness testified to the Pension Bureau in 1904: "He is nervous . . . [and] very excitable. Often afraid he is going to die and often begging some one to remain with him . . . The family and myself often sit by him quieting him and soothing him . . . He has a wild look from the eyes. Talks constantly and incoherently. The Doctor has often been compelled to stay with him all night." Needless to say, the burden of caring for James B. Farr fell mainly on his wife, Elizabeth, who testified as follows: "My life is one of constant watchfullness and care over him day and night, never leaving him or permitting him to go out of my sight without being with him or having someone else with him. I attend to all business matters and household purchases. We are constantly on the alert to prevent any noise or exciting causes from troubling him. [Even] with the best of care and watchfullness, he frequently becomes very violent. [We are afraid] he [will] relapse into a violent condition and we would again be compelled to send him to the Insane Asylum."[29]

Other accounts of Union veterans include similar stories of wives, who would stay up all hours of the night, talking with, reading to, or otherwise caring for and tending to their husbands. Perhaps there was a strength and purpose to these women, which did not need to be expressed in the form of radical individualism, gender wars, or a struggle against a white male patriarchy. In his memoirs written in 1905, one Union veteran described his mother in a way that may have been representative of the wives and mothers of most Civil War veterans: "She was made of strong stern enduring stuff . . . It required a mighty load to break her down." To characterize the circumstances of newly acquired responsibility for Civil War era women, resulting from the misery and misfortune of beloved husbands and menfolk, as in some sense empowering or "enabling" would be a distortion of these women's true feelings and devotion to family.[30]

In conclusion, the study of the Civil War should never be narrow and predictable, and limited simply to the repeated description of battles and campaigns, or the continued analysis and evaluation of famous generals and political leaders. Hence, it has been and will continue to be a positive development for social historians, with their interest in community and gender, to extend their focus to the Civil War. However, this new enthusiasm for social history and the study of the common man should be tempered by returning to Clausewitz's vision and understanding of war. As Clausewitz argued, battle is to warfare what cash payment is to finance: regardless of how complex the relationship between the two parties, and regardless of how rarely settlements may actually occur, they can never be entirely absent. Finding that unbridled violence lies at the core of war, Clausewitz concluded: "It is inherent in the very concept of war that everything that occurs *must originally derive from combat* . . . Fighting is the central military act; all other activities merely support it . . . direct annihilation of the enemy's forces must always be the *dominant consideration.* We simply want to establish this dominance of the destructive principle."[31]

We should never lose sight of the fact that war is about violence and battle, and the effect these phenomena have on the men called upon to take up arms, as well as on their families and communities. This is the irreducible bottom line in warfare. We must also recognize, consequently, that warfare has internal dynamics and a driving force all its own, and is not a mere continuation of society and social forces as one finds these in peacetime. While we should welcome the efforts of historians to study the common people in the Civil War—soldiers and their families—it is important that we attempt to understand these people on their own terms, and to discover what their trials, disappointments, and triumphs were during their experience of war. We should avoid superimposing our own agendas, preferences, and beliefs because the result will be a distorted and self-serving picture, rather than a revelation of truth.

Notes

1. Carl von Clausewitz, *On War,* edited and translated by Michael Howard and Peter Paret (orig., 1832; Princeton: Princeton University Press, 1976), 610.

2. *Ibid.*, 75, 77, 231, 228.

3. *Ibid.*, 583, 603, 592, and 219.

4. Fritz Stern, ed., *The Varieties of History: from Voltaire to the Present* (New York: Random House, 1973), 19; Gertrude Himmelfarb, *The New History and the Old: Critical Essays and Reappraisals* (Cambridge: Harvard University Press, 1987), 1–21;

Theodore S. Hamerow, *Reflections on History and Historians* (Madison: Univ. of Wisconsin Press, 1987), 162–204.

5. Hamerow, *Reflections*, 165.

6. Ira Berlin, Barbara J. Fields, Steven F. Miller, Joseph P. Reidy, and Leslie S. Rowland, *Slaves No More: Three Essays on Emancipation and The Civil War* (New York: Cambridge University Press, 1992), 146 (policy), 6 [watch and wait], x [ordinary blacks as "redesigning their lives"]; see also xiv [slaves accomplished their own liberation and shaped the destiny of the nation], 4 [role of slaves as decisive], 6 [the weak led the powerful], 69 [black soldiers as playing a conspicuous role in Union victory], 121 [blacks as using military service to become agents of change]; Joseph T. Glatthaar, *Forged in Battle: The Civil War Alliance of Black Soldiers and White Officers* (New York: Meridian Books, 1991), 2 [self-emancipation], 3, 71 [commitment of blacks to military service], 168 [vision of black troops]; Barbara J. Fields, *Slavery and Freedom on the Middle Ground: Maryland During the Nineteenth Century* (New Haven: Yale Univ. Press, 1985). See also Ira Berlin, Barbara J. Fields, Thavolia Glymph, Joseph P. Reidy, and Leslie S. Rowland, ed., *Freedom: A Documentary History of Emancipation, 1861–1867: Selected from the Holdings of the National Archives of the United States*, vol. 1: *The Destruction of Slavery* (New York, Cambridge (UK); New York: Cambridge University Press, 1985); Ira Berlin, Joseph P. Reidy, and Leslie S. Rowland, *Freedom: A Documentary History of Emancipation, 1861–1867, Selected from the Holdings of the National Archives of the United States*. Series II: *The Black Military Experience* (New York, Cambridge (UK): Cambridge University Press, 1982); Leeann Whites, "The Civil War as a Crisis in Gender," in Catherine Clinton and Nina Silber (eds.), *Divided Houses: Gender and the Civil War* (New York: Oxford University Press, 1992), 11–13.

7. Reid Mitchell, *The Vacant Chair: The Northern Soldier Leaves Home* (New York: Oxford University Press, 1993), 25, 31, 72 [no purely masculine world], 75 [incomplete], 81–82 [women's tasks], 87 [domestic sphere].

8. Ibid., xiii [object]; 11 [masculinity]; 13 [process]; 14 [domestic sphere].

9. Clausewitz, *On War*, 77, 231.

10. George E. Ranney, "Reminiscences of an Army Surgeon," in *War Papers Read before the Michigan Commandery of the Military Order of the Loyal Legion of the United States*, vol. 2 (Detroit: James H. Stone & Company, 1898), p. 189.

11. Wladimir Krzyzanowski, *The Memoirs of Wladimir Krzyzanowski*, translated by Stanley J. Pula, edited by James S. Pula (San Francisco: R & E Research Associates, 1978), 49; Letter of Phineas A. Hager, Sergeant in Company B of the 19th Michigan Infantry, to his wife, March 5, 1864 (Bentley Historical Library), [wickedness].

12. Earl J. Hess, *The Union Soldier in Battle: Enduring the Ordeal of Combat* (Lawrence: University Press of Kansas, 1997), 4 [grisly shock to brutal reality], 20 [these veterans].

13. In general, see Gerald Linderman, *Embattled Courage: The Experience of Combat in the American Civil War* (New York: Free Press, 1987).

14. Elbridge J. Copp, *Reminiscences of the War of the Rebellion, 1861–1865* (Nashua, N.H.: Telegraph Publishing Co., 1911), 242 [Dante].

15. Rufus W. Jacklin, *Records of the Military Order of the Loyal Legion of the United States, Michigan Commandery,* Bentley Historical Library [1st Co. Sharpshooters, 16th Michigan Infantry], 14; Robert S. Robertson, *Diary of the War,* edited by Charles N. and Rosemary Walker (Fort Wayne, Ind.: Allen County-Fort Wayne Historical Society, 1965) [93rd NY Infantry], 23–24; Curtis Buck, Buck Family Papers, Bentley Historical Library [Battery B, First Light Artillery]; letter of May 23, 1864, from Rome, Georgia; Eli Augustus Griffin Papers, Bentley Historical Library [Captain, Co. A, 6th Michigan Infantry; Major and Lt.-Col., 19th Michigan Infantry], diary entry of May 3, 1864, describing Chickamauga; Franklin H. Bailey Papers, Bentley Historical Library [Co. D, 12th Michigan; Co. E, 4th Cavalry], letter of April 8, 1862 describing Shiloh.

16. 1865, hanging in the Metropolitan Museum of Art, an image of this painting can be seen at http://www.metmuseum.org/TOAH/hd/homr/hob_67.187.131.htm.

17. Hess, *The Union Soldier in Battle.*

18. *Ibid.*, ix and 195 [resisted being passive victims], 4 [real soldiers], 47 [coherent vision], 74 [test role in war], 93 [courage, saving the Union], 156 [bitterness], 193 [failed emotional challenge], 195 [emotional response].

19. A. J. Glass and M. S. Mullins, eds., *Neuropsychiatry in World War II,* vol. 2, *Overseas Theatres* (Washington, D.C.: GPO, 1973), 135, 64 [symptoms]; Roy R. Grinker, "Brief Psychotherapy in War Neuroses," in *War Psychiatry,* Proceedings of the Second Brief Psychotherapy Council, Chicago, Ill., Jan. 1944 (Chicago: Institute for Psychoanalysis), 11 [cracked record].

20. Affidavits of James H. Kinnear, Feb. 11, 1890 [Chester Station; morbidly melancholy; bunk mates]; Alice S. Irwin, Jan. 21, 1890 [bar doors; whisperings; company cut up]; Daniel H. Demaree, Nov. 15, 1888 [jail; poorhouse], Dr. J.H. Reynolds, Nov. 16, 1888 [excitable; firing of gun; frightened]; Walter H. McElroy, Nov. 17, 1888 [Battle of Bermuda Hundred]; Lucinda Thorn, Nov. 24, 1888 [abusive to wife]; Henry Jines, Nov. 27, 1888 [threats against wife]; Jacob W. Jines, Nov. 27, 1888 [horses; wife], federal pension file of Dixon A. Irwin [B & D, 13 Ind. Inf.], National Archives. In general, see Eric T. Dean, Jr., *Shook Over Hell: Post-Traumatic Stress Disorder, Vietnam, and the Civil War* (Cambridge: Harvard University Press, 1997).

21. Affidavits of Dr. John N. Parrish, October 6, 1886 (after return from service, dispirited, gloomy, and "laboring under false impressions"]; Francis Miller, October 5, 1886 (housekeeper) [after return home from army in 1866, "out of his mind" and would walk around all the time, as though he didn't know what he was doing]; John H. Miller, October 5, 1886 [Guile would walk the streets in a kind of aimless way]; James Guile (guardian; brother), October 5, 1886 [incident at Atlanta; Guile gloomy, dispirited, shunned relatives, acted wild, wanted to kill somebody]; John Thrasher, April 5, 1995 [wild, obsessed with war]; Anna Comstock, April 5, 1895 (niece) [wanders at night with hatchet]; John Shelk, April 5, 1895 [shaking]; William J. Lester,

9/1/87 (comrade) [addled]; George W. Young, October 6, 1886 (lawyer): [talk flighty-like]; Ann Eliza Guile, October 6, 1886 (sister-in-law) [furlough], federal pension file of William H. Guile [I 63 and H 128 Ind. Inf.], National Archives.

22. Inquest papers for Hickman Dean, commitment #9376, August 16, 1893, Indiana Hospital for the Insane, Indiana State Archives ["imagined that some one was trying to kill him"]; Commitment ledger for Michael Decamp, commitment #8454, April 2, 1890, Indiana Hospital for the Insane, Indiana State Archives [Delusion that some one is after him trying to kill him]; Commitment ledger for Calvin Fishel, commitment #3041, Dec. 12, 1874, Indiana Hospital for the Insane, Indiana State Archives ["Fearful that persons will kill him"]; Inquest papers for Jacob Fink, commitment #5320, March 6, 1882, Indiana Hospital for the Insane, Indiana State Archives [fortified house]; Inquest papers for Benjamin F. Ford, commitment #5995, Nov. 2, 1883, Indiana Hospital for the Insane, Indiana State Archives ["He thinks there is some person after him to kill him"]; Inquest papers for Leo. C. Griffith, commitment #8308, Sept. 15, 1889, Indiana Hospital for the Insane, Indiana State Archives ["In dread of being killed"]; Inquest papers for Titus C. Jones, 1892, Indiana Hospital for the Insane, Indiana State Archives ["afraid of being captured and murdered"]; Inquest papers for Henry Barnhart, commitment #7906, July 12, 1888, Indiana Hospital for the Insane, Indiana State Archives [fire arms in readiness]; Affidavit of Harvey R. Benshan, June 5, 1894, federal pension record of John A. Cundiff [H 99 Ind. Inf.], National Archives; Letter from Almena C. Gaither, Jan. 19, 1903, federal pension file of Allen E. Wiley [C 54 Ind. Inf.], National Archives ["thought someone was going to shoot him"]; Inquest papers of Elias Hammon, commitment #6020, Nov. 17, 1883, Indiana Hospital for the Insane, Indiana State Archives [calls for gun]; Inquest papers of Demarcus L. Hedges, commitment #9193, Aug. 26, 1892, Indiana Hospital for the Insane, Indiana State Archives [kill and bury him; Vicksburg].

23. Catherine Clinton and Nina Silber, eds., *Divided Houses: Gender and the Civil War* (New York: Oxford University Press, 1992).

24. Kristie Ross, "Arranging a Doll's House: Refined Women as Union Nurses," in *Divided Houses,* 107, 110, 105.

25. Drew Gilpin Faust, "Altars of Sacrifice: Confederate Women and the Narratives of War," in *Divided Houses,* 172, 175, 195.

26. Willard Waller, *The Veteran Comes Back* (New York: The Dryden Press, 1944), 37.

27. Records of the Illinois State Hospital for the Insane at Jacksonville, Illinois State Archives, Springfield, Illinois: Lydia A. Davis, commitment #1625, June, 24, 1862 [husband]; Millicent C. Loar, commitment #1653, July 11, 1862 [brothers]; Maria Walsh, commitment #1943, February 10, 1864 [violent]. Indiana Hospital for the Insane, Indiana State Archives: Mary Jane Moore, commitment #2546, Sept. 2, 1863 [death of son]; Mary Jane Sage, commitment #2552, Sept. 9, 1863 [grieves]; Emily Kuipe, commitment #2681, April 29, 1864 [death of brother]; Cordelia McGannon, commitment #2703, June 18, 1864 [anxiety about husband]; Lucinda St. John, commitment #2706, June 24, 1864 [absence of husband]; Anna R. Johnson, commitment

#2759, Sept. 26, 1864 [husband wounded]; Hester A. Taylor, commitment #3449, Nov. 6, 1867 [husband died in army]. Commitment record of Martha Harbin, March 28, 1864, South Carolina State Hospital [husband killed in army]. Columbus (Ohio) Hospital for the Insane: Mary Ann McDonald, commitment #4422, May 31, 1865 [supposed cause the death of son in army]; Mary Hubbard, commitment #4263, June 15, 1864 [Cause—loss of brother in the Army]; Ellen Strayer, commitment #4258, June 13, 1864 [Cause loss of son in army].

28. Memoirs of James L. Cooper, Confederate Collection, Tennessee State Archives, Nashville, Tenn., 43.

29. Affidavits of J. J. Johnson, December 6, 1888 [dizzy]; medical board at Martinsville, Indiana, June 4, 1881 [mental faculties impaired]; Idem, May 29, 1889 [nervous]; Idem, Nov. 15, 1899 [melancholy, sleepless, loss of memory]; Idem, April 20, 1904 [description of wound]; Josiah L. Bunton, Jan. 2, 1904 [afraid he is going to die] Elizabeth Farr, Jan. 2, 1904 [wife], federal pension record of James B. Farr (H 33 Ind. Inf.), National Archives, Washington, D.C.

30. Memoirs of DeWitt C. Goodrich, Indiana Historical Society, Indianapolis, Indiana, 38.

31. *On War*, 95 [combat]; 97 [destruction], 338 [violence; emphasis original].

5
Civil War History and
the Myth of the Lost Cause
National Consequences of Civil War Battle

Alan T. Nolan

I am dissatisfied with the status in the popular mind and popular memory of what is commonly thought of as the history of the Civil War. I seem to be dissatisfied a lot these days; I seem to be the Grinch of Gettysburg. I think that what I consider to be bad history is largely a consequence of the work of the Virginia cult, led by General Jubal Early immediately postwar, in the Southern Historical Society Papers. Those people were clever. You have to give them credit; they were bright. They realized that the memory of the war was up for grabs right after Appomattox, and very early they got busy creating a Southern rationale of the war. Douglas Freeman, Avery Craven, and Clifford Dowdy in modern times added to that. Then of course there was popular entertainment like *Gone with the Wind,* which is an incredibly good story but very bad history, but which lingers in the popular mind both in the South and in the North. And then the whole Lost Cause Myth—which as I say is a Southern rationale—is essentially a romantic take on the war.* I think that kind of romanticism and distortion is not an acceptable public memory of the defining event of American history.

As we all know, there was nothing cute or winsome about the Civil War. We who are very interested in it have an obligation to understand the war as history. I intend to identify and comment on some of the elements of the Lost Cause Myth.

The single most profound element of the Southern myth concerns the relationship of slavery to the war. Now that, you would think, is a

---

* The term "Lost Cause" as applied to the South in the Civil War era was popularized by Edward A. Pollard's *The Lost Cause,* an interpretative history of the war from a Southern viewpoint, published in 1866. See James M. McPherson's definition of "myth" in the Introduction to this volume.—Editor's note.

historical question. But it becomes more and more apparent that it becomes not so much a historical question as a psychological question. I think that psychologists would say that the Southern neo-Confederates and the Southern partisans of the war are really in denial when it comes to the relationship of slavery to the war. It seems perfectly obvious to me that the Southern states seceded in order to protect slavery. In that sense the war was caused by slavery; but in any event, the important thing is that the premise of the Confederacy and the reason that the Confederates withdrew and then fought was to protect the institution of slavery. Period. You don't have to say "there were a lot of other things:" that's baloney. But for slavery and the South's commitment to it and intent to protect it, there would not have been secession and therefore there would not have been a civil war.

Now we all know that at the very beginning of the constitutional United States, slavery was somewhat mooted; but it did expressly come up, as some books, such as *Founding Brothers*[1] and the recent Adams biography,[2] make perfectly plain. The subject of slavery came up at the Constitutional convention and was set aside. I recommend to all of you one of the most illuminating books I have ever read about American history, and I have been reading about American history for a long time: Stanford's professor Fehrenbacher, who died recently, wrote a book called *The Slaveholding Republic.*[3] It is an excellent treatment of what slavery did to the first seventy or so years of American history. But we also know of course that by 1820 the admission of Missouri to the Union was an extremely hot issue in the minds of people, North and South. Then we had the Mexican war, dominated by the issue of whether its purpose was to acquire territory for slavery. That is surely what Lincoln thought. The whole issue of new states during the period prior to the Civil War was always complicated by the question of whether they would be slave or free. We also know that on the eve of the war, intensifying as the years went by, the question of fugitive slaves occupied national attention. A lot of Northern people had been able to say, "Slavery is not my problem; it's the problem of those folks down South." But then the Southerners decided to make the Northerners catch slaves. That changed the dynamic of slavery a great deal. As a legal proposition, if the government can turn to you and say, "You're going to be a slave catcher," that of course would be anathema to the Northern people. We also know about the gag rule, which required the tabling of petitions to Congress from people wanting to abolish slavery.

And then came the Compromise of 1850, which entirely concerned the legal status of slavery.

I think the most significant evidence in favor of treating the role of slavery as a historical question as opposed to a psychological question—the most profound evidence—is the pre-Sumter compromise proposals. The Crittenden and the various other unsuccessful compromise efforts all concerned *the legal status of slavery*. So unless these men, North and South, were out of their minds, you have to acknowledge that the issue underlying secession was the South's demand for the protection of slavery. I refer you to the Confederate constitution, Article One, Section Nine, and Article Four—both of which provide for slave property being protected.

Mr. Fehrenbacher, along with much recent scholarship, points out the fact that almost every issue debated by the national legislature—the location of the transcontinental railroad, for example—boiled down to the question of "slave or free?" And Fehrenbacher shows that the issue of "slave or free" captured the imaginations of Southern people and Northern people.

I also recommend *The Apostles of Disunion,* by Charles B. Dew,[4] published by the University of Virginia Press. It refers to the fact that when the Deep South states seceded, they appointed delegates to go to the other slave states to try to talk their legislatures into also seceding. This book publishes what those agents to the not-yet-seceding states were concerned with: they were concerned entirely with protecting slavery.

But in spite of all that, the neo-Confederates tell us that slavery had nothing to do with the war. It's the most monstrous kind of dissembling I have ever been aware of. Perhaps you have seen that this year a very good Civil War magazine, *North and South,* has been concerning itself with some of the cosmic issues about the war—such as, What caused it? What was the role of slavery? What was slavery's relationship to the war? James McPherson of Princeton wrote an article pointing out the involvement of slavery with secession and the establishment of the Confederacy. He argues, as I argue, and as I think any serious historian would argue, that the South seceded to protect the institution of slavery. That was the foundation of the Confederacy; that was what it was all about. The magazine published that article, and then the next six issues contained long letters from Southerners saying that slavery had nothing to do with the war. It's really an economical way to run a magazine: publish one article critical of the myth and then sit back and publish the neo-Confederate responses.

So that is the fundamental issue in any kind of discussion of the Lost Cause Myth: the fact that slavery caused secession and therefore caused the war.

Another interesting element of the Lost Cause Myth is the bête noir of the whole thing, the abolitionist. Having carved away slavery as the cause of the war, the Southern mythmakers have to have some cause for it; so they say it happened because of those damned abolitionists. They are described in Southern literature as fanatics who manufactured a problem that didn't really exist. Those crazy abolitionists just made it all up out of whole cloth—made everybody unhappy and brought on the war. The trouble with the abolitionists, of course, was that they were *right* at the time and most of the other people were wrong. And to harbor in American history the negative image of the abolitionists, which still lingers in contemporary books about the war, and to harbor an anti-abolitionist feeling—is historically outrageous.

Another of the principles of the Lost Cause is that the South would have voluntarily given slavery up; it was simply a question of time. Historically, that is clearly not so. In the first place, you look to what the Southern states were doing to civil liberties. Constitutionally, they were locking themselves into slavery. In Virginia in 1855, the criminal code provided that if you said slavery was wrong—if you *spoke* something critical of slavery—it was a felony, and you would go to the penitentiary. And in all the Southern states that seceded, the Constitutional concepts of freedom of speech, freedom of the press, and freedom of assembly were abandoned completely in the interest of making sure that slavery was not subject to public criticism. So the idea that the South would have given slavery up, given the fact that they were locking it into their legal system every year through their legislatures, seems to me to be far-fetched.

Furthermore, you have to look historically at the filibustering about buying Cuba. There was a tremendous movement in that direction. The last big effort was made by Buchanan, who was of course a dough-faced Democrat.** The idea was to acquire Cuba so that it would be United States slave territory. There was also a lot of filibustering about Central America. The idea was that there would be a North America–Central

** "Dough-face" was a term given to Northerners who supported the South and/or were sympathetic to slavery.—Editor's note.

America–South America slave kingdom. So the South was not talking about giving up slavery; they were constantly talking about expanding it.

Another area of myth—and one of the areas that I think is most annoying—is the nature of the slaves themselves. The Lost Cause invented two interesting stereotypes. One was the "happy darky." You saw this time and time again in the movie *Gone with the Wind.* Hattie McDaniel won an Academy Award for being a happy slave. The "happy darky" stereotype passed through Walt Disney down to our own day. The other invention was the so-called "faithful slave"—the slave that was loyal to and loving of his or her indulgent master.*** And of course those concepts are absurd. We know that the major logistical problem of the Federal armies—you can read this in the *Official Records* (OR) any day[5]—was that as the Federal lines advanced into the South, thousands of slaves left the plantations and took refuge in the Federal lines although they were not well-treated as they did so. The idea of these slaves wanting to hang around Old Massa and being so happy to be slaves seems to me to be an insult. We also know, of course, that approximately 180,000 former slaves volunteered to fight in the Federal army. And there are many, many accounts of Federal soldiers—escaped prisoners, or in some other way being at large in the Confederacy and being pursued—being helped by people in the South. Invariably the people who looked after them—the people who fed them and took care of them and secreted them—were slaves. For example, John Kellog of the Sixth Wisconsin of the Iron Brigade wrote a wonderful book called *Capture and Escape.*[6] The book recounts his trying to get back to Federal lines across a long distance. If he saw a black person, he knew he could count on that person: that black person, male or female, would take care of him. So the idea of good ole Massa and black people being so keen on slavery is, I think, an insulting myth.

Another characteristic of the Lost Cause Myth is the idea of a North/South nationalistic cultural difference, by which Southerners, à la Sir Walter Scott, created the idea of the Southern people being descendents of the Normans, those allegedly very superior Europeans who went over to Great Britain and conquered the Anglo-Saxon tribes. The Anglo-Saxon tribes' descendents were the Yankees, according to the Southern

*** Those who have seen the movie *Gods and Generals* will be familiar with a contemporary version of what Mr. Nolan refers to here.—Editor's note.

mythology. But Southerners were descendents of the Normans, who were claimed to be extraordinary people.

Another aspect of the Lost Cause is the rationalizing of the military loss. There was a general Southern theory, which was simply playing with semantics, to the effect that the South was not *defeated* in the war; it was somehow "overcome." James McPherson reminds us that the South was not only invaded and conquered, it was utterly destroyed.[7] By 1865, Union forces had destroyed two-thirds of the assessed value of Southern wealth, two-fifths of the South's livestock, and one-quarter—one out of every four—of the white men between the ages of twenty and forty. More than half of the farm machinery was ruined and the damage to railroads and industry was incalculable. Southern wealth was decreased by 60 percent. Now the fact is that in 1864 and 1865, Northern armies roamed at will through most of the South. The stories that you read about people wondering what to do about Sherman in the Carolinas show that there was simply nothing available to oppose that Federal army. So the idea that somehow the South was not defeated has no basis.

Of course one of the corollaries of that idea concerns poor old Lieutenant General James Longstreet. The theory is that "Longstreet lost it at Gettysburg." I think that the faulting of Longstreet at Gettysburg is not historically sound, in the first place. But the idea that Gettysburg decided the war is also not historically sound. Obviously that notion overlooks the March to the Sea, it overlooks Missionary Ridge—it overlooks many things that happened after July of 1863. Two years ago I was down in Richmond one weekend. I picked up a newspaper, and in it was a kind of cute account of an old Confederate veterans' home. The story was about these old veterans sitting around on the porch and a newspaper boy went by talking about Pearl Harbor. He brought a newspaper up to them and one of the old men said, "By God, if Longstreet hadn't been late at Gettysburg this never would have happened!"

Another aspect of the Myth of the Lost Cause is the idealization of the home front. The Moonlight and Magnolias society features very largely in the Myth, and it suggests that everybody was very happy about what was taking place in the war. But the fact is that the conflict and discord in the Confederacy made the Yankee conflict and discord look like peanuts. Bell Wiley—a great man and a great Southern historian, who grew up in the South and taught in the South—writes that strife was the evil genius of the Confederacy, and nothing that the Confederacy did was unmarked by that strife.

There's also an idealization of the Confederate soldier in the Lost Cause Myth. By the way, I am not denigrating the Confederate soldier. The greatest victim of the Confederate hoax was the Confederate soldier. According to the myth, he was faithful and law-abiding. Now the facts are to the contrary. It is amazing to me how many writers quote Lee's orders. When he went into Maryland, Lee did issue orders to his troops proscribing their behavior; and he also did that when they went into Pennsylvania. Writers quote those orders, but they do not then pay attention to what in fact the soldiers did. For example, in Maryland, after issuing those orders, Lee is in constant correspondence with Jefferson Davis about the misbehavior of his troops. I suggest that you look in the OR: early in the Antietam volume, read Lee's communications about how he could not control his soldiers. They were marauding all over Maryland, and the same was true of course in Pennsylvania during the Gettysburg campaign. In that campaign, among the things that the Confederate soldiers were doing in Pennsylvania was seizing and selling into slavery free blacks. So the conduct of the idealized Confederate soldier, faithful and law-abiding, is a myth. They were no worse or better than Federal soldiers. Soldiers are soldiers, unfortunately. Confederate soldiers behaved themselves like soldiers, which meant that a certain percentage of them were by no means law-abiding and by no means faithful.

Another of the characteristics of the Lost Cause is what I call "The Saints Go Marching In." This has to do with Lee and Jackson, who are surely mythological figures. Lee, I have written about, taking each of the myths about him and trying to respond to them. But the Jackson myth is also a powerful one. We are told how wonderfully religious he was— which seems to me to be somewhat irrelevant to the Civil War. There's no doubt that he was an effective soldier. But he was also like Cromwell among the Irish, killing people for the glory of God. And I think that despite the successful mythology that the Confederate publicists created about these two supermen—Lee being a very angelic person, and Jackson being a sweetheart—the fact is that Jackson was an almost impossible person. If he disagreed with you, he'd place you under arrest. Even more outrageous is the apotheosis of Nathan Bedford Forrest. He has become the third member of this trinity of Confederate leaders. Undisputedly,

Forrest's personal fortune came from slave trading. It is also undisputed that after the war he was one of the founders and leaders of the Ku Klux Klan. It seems to me that he would make a very strange American hero. But in the Lost Cause mythology he has become one.

Now what about the legacy of the Lost Cause? Its legacy to history is such clichés as "both sides were right," which makes me want to lose my hardtack. Breaking up the United States of America in 1860 was not in anybody's interest. There would have been totally negative consequences. Had the North and South separated, there would have been constant strife about access to the Mississippi River; there would have been constant strife about who was going to have what from the territories of the United States government. I know that a relative minority of Southern people owned slaves and therefore had a direct property stake in slavery, but except for slave owners, I can't see any class of Southern people who would have been benefited by breaking up the United States.

The Lost Cause leads to a cliché that Northern people have to some extent bought, that somehow the war was like the Harvard-Yale football game. Nothing was really at stake, just a bunch of nice guys with the South and a bunch of nice guys with the North and they had this war. For God's sake, there was a tremendous amount at stake! The integrity of the United States government, territorial and political; human freedom—as imperfect as the war was in terms of the issue of slavery—at bottom, human freedom was involved in the Civil War, and both sides were not right about that, not by a damn sight.

So the legacy of the Lost Cause to history has been distortion. It is a kind of cartoon of the facts of history. The political legacy is a little more complicated. I don't think there is any question about the fact that reunion between the Northern states and the Southern states was facilitated by the North's beginning to accept the Southern rationale. The North was incredibly racist, as was the South: racism was the common denominator of the North and South. As a matter of fact, racism was the express price that was paid for reunion. The Hays-Tilden compromise, which elected Hays as president of the United States on the agreement to remove Federal troops from the South, was a quid pro quo. Really, the only protection the freedmen had had was the Federal troops. When they were gone, the term used was that the Southern states were "redeemed;" that is, blacks no longer participated in politics and were violently treated. So the reunion—the white men's reunion, the racist white men's reunion between the North and the South—was clearly facilitated, I think, by the Lost Cause Myth and by the fact that the North became a sort of partner in that, accepting that myth.

The continuing social and political legacy of the historical distortion has been racism. The nation went into Jim Crow until 1954 as a result of the North's buying the Southern mythology about the happy darky, that foolish character—a clown rather than a human being—who really liked being a slave and was crazy about his owner. Buying the myth meant that we didn't challenge its implications and its consequences.

You know the Irish are at their best when they've got somebody to be mad at, and, being Irish, I am arguing for truth and accuracy about Civil War history rather than the clichés we have been brought up with. I learned in grade school that the war was not about slavery. I remember seeing a movie when I was in grade school, with Shirley Temple and Bill Robinson dancing and being so cute to each other—she being a Southern girl and he being a slave. We have a kind of brainwashed concept that the war was just sort of a sporting event, and everybody was swell. And I am saying that I resent that desperately. I feel that way because it is so damned unhistorical.

Civil War history ought to be about being accurate.

Notes

1. Joseph J. Ellis, *Founding Brothers: The Revolutionary Generation* (New York: Vintage Books, 2000).

2. David McCullough, *John Adams* (New York: Touchstone, 2001).

3. Don E. Fehrenbacher, *The Slaveholding Republic: An Account of the United States Government's Relations to Slavery* (New York: Oxford University Press, 2001).

4. Charles B. Dew, *The Apostles of Disunion: Southern Secession Commissioners and the Causes of the Civil War* (Charlottesville: The University of Virginia Press, 2002).

5. The War of the Rebellion: A Compilation of the Official Records of the Union and Confederate Armies (Washington, D.C.: Government Printing Office, 1889).

6. John Azor Kellog, *Capture and Escape: A Narrative of Army and Prison Life* (Madison: Wisconsin History Commission, 1908).

7. See James M. McPherson, *Drawn with the Sword: Reflections on the Civil War* (New York: Oxford University Press, 1996), chapter 5 and p. 240. (Editor's note.)

6
Numbers

Kent Gramm

At first glance, cold numbers seem to say nothing about human experience; in fact, they can reveal unexpected things about the nature and consequences of Civil War battle. Numbers are windows to how the Civil War generation might have felt in battle, and after battle. They are clues to the actual nature of the war, resistant to misconceptions and popular mythology. Numbers can tell us something about who we are.

What scared a soldier on a Civil War battlefield? Artillery. Do the numbers show that artillery was worth being afraid of? If not, were Civil War soldiers oversensitive civilians whose intermittent, relatively brief battlefield experiences never hardened them to war?

Were Civil War soldiers "efficient"? Did they do at all well what modern soldiers are trained to do—kill? War is about fighting, and fighting means killing, as William T. Sherman pointed out. Was the Civil War soldier a good killer? What can we learn about his beliefs and his training? Numbers give us good glimpses into the answers.

We know the civilians of that distant era were a sentimental people, judging by their literature, music, and popular tastes. Did they exaggerate the experience of war, and would we have been less affected by the war than they were?

Was the nature of the Civil War, stripped to its essentials, the remorseless and irresistible pounding of a strong contestant on a weak one, with its outcome a foregone conclusion? Has the drama of the Civil War been exaggerated, therefore; and should we understand the war in less grandiose terms, reducing its mythic footprint to a more pedestrian size?

Finally, have the war's effects diffused, after 145 years, into the tumultuous atmosphere of the postmodern world? If sentimental nineteenth-century people exaggerated their experience, both on the battlefield and on the home front, then the war belongs in a museum: let's give it a rest. But if the experience of that war would have also overwhelmed us today, then it was truly a shocking and transformative experience, and we twenty-first-century Americans would be wise to try to assess our influences, in order to understand what is important to us and what we are capable of.

10

This simple, basic number is not only the cornerstone of our decimal system, it is the foundation of our understanding the true impact of the Civil War—the effects on its participants as well as the consequences to us.

Consider the consequences of September 11, 2001. In one day, we found ourselves in a different world. The shock and mental oppression of the event could hardly be understood by anyone outside the country. Economically, the event struck two mainstays of the American structure: transportation (especially our travel) and defense. It was uncertain whether Americans would indeed go back to the malls, continue to plan vacations, and buy major items like homes and cars. Before long, we became involved in two overseas wars, which have now lasted longer than the Civil War, and will have vast consequences economically, geopolitically, and morally. At the time of this writing, the United States is debating the very nature of American freedom as the sometimes conflicting requirements of security and liberty battle it out, seemingly to the death. Our institutions, values, and future are at stake. All resulting from 9/11.

The number 10 puts the Civil War into perspective alongside September 11. The population today is roughly ten times the population of Civil War America. The 1860 Census put the figure at 31,443,321.[1] What this means is that to understand the magnitude of any loss of life in the 1860s, we need to multiply it by 10. A commonly agreed-upon rough figure for Civil War deaths is 650,000—or 6,500,000 in today's terms.

If you divide that figure by the four years of the Civil War—1460 days—you see where the Civil War fits in relative to 9/11. It was September 11, 2001, *every day, for four years*. Only worse: instead of about 3,000 people per day, it would be 4,452.

The Civil War generation did not exaggerate their experience. Judging by how we reacted to 9/11, we can see that the impact of the Civil War was, and no doubt continues to be, profound, vast, and incalculable. In America, particularly in the Southern states, where one in four military-aged men died, the fatality of the war equaled or surpassed the mortality of both world wars in Germany, France, England, and Russia.[2] The Civil War generation might have been sentimental, but if anything they were mentally tougher, rather than weaker, than we are. For four years they sustained and withstood an assault on American institutions, nationhood, and values—coming out not with a crippled remnant of the Spirit of '76 but with "a new birth of freedom."

Along these lines, it becomes clear that the institution of slavery in America was no inconsiderable presence. In today's terms, there would be 35 million slaves. The preoccupation of abolitionists and slaveholders and eventually of President Lincoln with this critical mass of people is easier to understand when put into contemporary terms. We can also get a sense of the difficulty of Reconstruction, when the status of such a number of people radically changed overnight. Should the 35 million be concentrated in one section of the country, we can understand why that section's economy did not recover for a hundred years, and why the problem of readjusting social relations is taking so long. Add the perennial factor of American race feeling, and it is a wonder that the country has done as well as it has, even as it remains clear that the issue's magnitude is still with us.

Finally, contemporary Americans are generally unaware of the physical consequences of battle in Vietnam and the Gulf wars, as the disabled veterans of those conflicts are relatively few and out of sight. In the Civil War, one of three deaths was caused by battle. Let us use the conservative figure of 200,000. How many amputees per death were incurred cannot be determined. About 4.5 soldiers were wounded for every man killed. Federal data show that 71 percent of wounds treated during the war were to legs, feet, arms, or hands.[3] It is safe to say that a high percentage of these wounds received amputation. Let us conservatively estimate 20 percent. These speculative figures would yield 1,278,000 amputees in today's terms. The number is probably far too low, but even at that number, these men would hardly be invisible.

In today's terms the Civil War produced not only 6½ million deaths, and at least 1¼ million amputees, but 30 million veterans, 35 million ex-slaves, and (assuming the average family having two adults and four children) 32.5 million people who lost a member of their immediate family.

In other words, over 105 million of our people would be *directly* affected by the war. That is nearly a third of our population. It is no wonder that the Civil War is still with us.

15–20 Percent

At Brawner Farm the day before the second battle at Manassas, two Federal brigades engaged a reinforced Confederate division at 70 yards maximum for approximately 1½ hours. Union casualties were, in killed/ wounded/missing, 1,025; Confederate losses totaled 1,250.[4] The losses were approximately one-third on the Federal side. Men firing one round every 30 seconds at close to point-blank range should have been able to kill each other off in a few minutes. Instead, they stood and fired, almost shoulder-to-shoulder in line of battle, until one side withdrew at a walk.

At Gettysburg on July 2, the 9th Massachusetts Battery retreated while firing, in the face of a Confederate infantry attack and under artillery fire. Standing with four of its six guns in a fence corner across from the Trostle house, the battery was assaulted not only from in front but on two sides, until eventually Confederate riflemen climbed atop caissons parked behind the guns and shot at the cannoneers, officers, and horses. One of the battery's survivors later said that he wondered how anyone survived. From a battle strength of 104, the 9th Battery suffered 7 killed and 4 mortally wounded.[5] Adding captured and wounded, the battery suffered a loss percentage of 27 percent[6]—surrounded on four sides by all or parts of three Confederate regiments. What are we to conclude about the intensity of Civil War combat? One might well ask how anyone survived.

The answer might be in the numbers 15–20 percent. In *Battle Tactics of the Civil War,* Paddy Griffith estimates a kill rate between regiments firing in line of battle at one to two men per minute.[7] Most Civil War soldiers were armed with Springfield or Enfield .58/.577-caliber rifled muskets that were accurate at 200 yards and could kill at 400 yards. (Many of the Federals at Brawner Farm were armed with less modern weapons, but they were accurate at the short range described.) How, then, could anyone survive the stand-up firefights that dominated Eastern battlefields until after Gettysburg?

According to Lt. Col. Dave Grossman in *On Killing: The Psychological Cost of Learning to Kill in War and Society,* "at least half of the soldiers in black-powder battles did not fire their weapons, and only a minute percentage of those who did fire aimed to kill the enemy with their fire."[8] That "minute percentage" was 15–20 percent.[9]

After the battle at Gettysburg, roughly 30,000 muskets were found on the field, of which 85 percent were loaded. Of those 30,000 muskets, 40 percent contained multiple loads.[10] Given the repetitive training that instilled the "load in nine counts" in soldiers, but omitted firing, we might expect soldiers to merely go through what they were trained to do (i.e., load), once the excitement of battle nullified other thoughts. But presumably soldiers obeyed other orders; why not "fire"? Although the figure of 15–20 percent is hard to prove, the U.S. Army believed it, and instituted a new training program that has reduced non-firing from the calculated 85 percent to 5 percent.[11] This training involves extensive target shooting, particularly at targets that pop up unexpectedly like humans, and in some cases look like humans, so that the tendency to follow orders overrides the reluctance to kill other people outright. You are trained to shoot at "targets," not "soldiers," and your reflexes have been conditioned to fire at moving, sudden shapes, rather than stationary targets looking like bull's-eyes. Not only did Civil War soldiers get little or no target practice (so as not to waste the ammunition that the present industrial capacity produces in abundance), they were not conditioned to fire in this way. Indeed, their weaponry did not permit them to: a long, heavy Springfield makes an M16 feel like a toy by comparison. But the toy can be brought up in an instant and can fire a Civil War company's number of bullets with one squeeze.

Learning that to kill does not come naturally to most people is a comfort. However, to balance this agreeable consideration, one must remember that regiments often did stand until sufficient killing was done—even if only one in five men were doing it. And there were always the exceptions that question the low figures. To mention a single instance: on the second day at Gettysburg the 1st Minnesota lost 68 percent of its men in a 15-minute firefight with an Alabama brigade. Some of the losses were sustained in attacking through an open field under artillery fire, and the Minnesotans confronted more than a single regiment's fire. Still, if Griffith's one or two men per minute figure was meant as an average, the 1st Minnesota should have lost 50, not the over 200 actually recorded.[12] The Alabamians seemed not to have read Griffith's and Grossman's books.

Nevertheless, it would appear to be true overall that only a minority in each regiment did the killing. This means that a few minutes of line-of-battle firing looked somewhat more like a present-day reenactment than a person might think—except that the casualties stayed down. One by one, as a kind of time-lapse event, regiments stood in the line of fire and stayed until they were overborne or outflanked or ordered back. The average Civil War soldier had a sense of pride, was reluctant to let his comrades down, stood up for his convictions, obeyed orders in the selective manner of a free American citizen, and did not approve of killing. A Civil War battlefield enabled him to maintain these values somewhat intact.

11 Percent

At Gettysburg, artillery accounted for 11 percent of wounds treated: shell fragments 9 percent, canister 1 percent, and solid shot 1 percent. This is a nice, firm figure that might tell us something about Civil War combat, but it might also tell us nothing, or even mislead us. We need another figure to put the 11 percent into context. That figure is the number of "Missing."

One could conclude from the 11 percent that about nine of ten wounds in battle were inflicted by small arms. One could then conclude that artillery made a lot of noise but presented an inordinate drain on an army and upon the manufacturing capacity of North and South, proportionate to its real effectiveness. On the other hand, a regiment might be willing to stand in front of another regiment firing for an hour, but would they stand in front of a battery? The soldiers, perhaps, were scared by more than a battery's noise.

It does not take much imagination to see the limitation in the 11 percent figure. How many of the wounds inflicted by artillery were treatable? Failing to find a thorough tabulation of wounds visible on soldiers buried on fields of battle, we must look for less direct evidence. The category of "Missing" refers primarily to the number of soldiers captured; however, the references by survivors to the effects of close range artillery fire—such as causing an entire front line to disappear, or giving a pinkish tint to the air—suggest that "Missing" can be understood in an all too literal sense.

Ideally, one could isolate the missing from Pickett's and Pettigrew's divisions on the afternoon of July 3 at Gettysburg, but the large numbers of captured attackers prevents this. Was there an attack on artillery during the war that did not result in many captures, for which we have somewhat reliable casualty data? A possibility is Hays and Avery's attack on East Cemetery Hill on the evening of July 2 at Gettysburg. Hays's brigade suffered 61 killed, 187 wounded, and 86 missing. Avery lost 92, 213, and 107. It is impossible to isolate the missing caused by the effects of artillery; however, the brigades advanced right up to and among the cannons, retreating before the advance of Union infantry. Carroll's Union brigade captured "an assortment" of Rebels,[13] but evidently not many, and certainly not 195. Wiedrich's and Rickett's batteries no doubt accounted for many of the missing with their loads and double loads of canister. Both Avery and Hays reported fewer killed than missing, which seems an unlikely outcome given the nature of that encounter, if "missing" meant only the men lost to capture.

Other indirect evidence exists on a large scale. It is generally understood that in the Civil War, the advantage lay with defense. However, the reason given for this—the increased accuracy and range of Civil War small arms since the development of close-order tactics—is probably wrong. The Springfield and Enfield rifle-muskets were a real workout to fire, not only because of their weight but also due to their punishing kick. Weapons fouled quickly. They were not easy to aim at ranges above 100 yards, because the distance had to be estimated and the weapon's sight adjusted accordingly. The big slug traveled slowly along an arc at such ranges. Therefore "it is difficult to find any evidence at all to support the suggestion that Civil War musketry was delivered at ranges much longer than those of Napoleonic times—"which means between 30 and 75 yards.[14] Furthermore, regiments were not taught to shoot. The 24th Michigan, one of the most effective regiments of the war, received only one target practice (during which three soldiers were wounded and one died of a heart attack)[15] before its first engagement and only one more between that engagement (Fredericksburg) and Gettysburg. All of these considerations cause one to doubt Grossman's figure of 85 percent non-firers during the war, because certainly these other factors were important. An infantry firefight was not only conducted by men who might have been reluctant to kill, but it was a slow-firing affair, with inaccurate shooters. When an efficient unit set to it, such as happened in the Cornfield at Antietam, the losses were quick and heavy: one of Hood's Texas regiments lost 82 percent of its members in twenty minutes.

But the Cornfield was not an example of what Civil War soldiers meant by a "defensive position." Such a position meant artillery, lots of it, advantageously placed. The prime example of defensive fighting is Fredericksburg. The attacking army lost 12,653 men, the defenders 5,309.[16] Burnside's divisions advanced over long stretches of open fields, most of the time beyond the 100-yard musket range, but well within the effective range of artillery, including canister. Burnside reported 1769 missing. (Artillery was not an important factor for the attacking side, because artillery charges were seldom made during in the war. Long-range artillery intended to support an attack did not have the effect of canister used by defenders when the attackers closed to within a few hundred yards.)

What may we conclude from these numerical speculations, regarding the nature of battle? Probably the strong impression made by artillery was justified by the actual carnage it wreaked among infantry, though this is not readily apparent from data related to wounds. The real terror of Civil War fighting was indeed artillery fire. Its punishing thunder was correctly associated with horrific wounds, and even with the prospect of being blown to nothing. To charge artillery at a walk took considerable courage, unit morale, and/or the strange abandon sometimes reported by soldiers: "[I] lost all fear and thought of Home and friends, and a Reckless don't care disposition Seemed to take possession of me. Then was two of our Company Shot down near me and Even their Shrieks and yells did not affect me in the least. This is the way I felt and I have heard other Soldiers Say the Same."[17]

To operate in such a terrifying and lethal environment as created by artillery, the soldier's invaluable—or suicidal, depending upon your point of view—ally was rage. This was understood from the earliest times; e.g., *The Iliad*: "Rage—Goddess, sing the rage of . . . Achilles."[18] And it remains true in modern times. An officer in the 6th Infantry during the North Africa campaign in World War II wrote, "A soldier is not effective until he has learned to hate. When he lives for one thing, to kill the enemy, he becomes of value."[19] Americans in North Africa did not reach parity with the Germans until, wrote war correspondent Ernie Pyle, "they had made the psychological transition from their normal belief that taking human life was sinful . . . [and] finally learned to hate."[20]

When General Eisenhower took Field Marshal Montgomery on a tour of Gettysburg and the two men looked across the field of Pickett's Charge toward the Union position, Eisenhower remarked that General Lee must have been so mad at Meade that he wanted to hit him with a brick. The same had to be true of many of the men who made the charge.

The "shot fired in anger" is a cliché whose truth has been forgotten. Nothing but the ecstatic wrath of battle explains the Angle at Spotsylvania, where Union and Confederate soldiers fought almost hand-to-hand, separated only by the width of a log breastwork: "The gray and blue coats with rifles in hand would spring on top of the breastworks, take deadly aim and fire, then fall dead in the trenches below."[21]

Those who wished could fail to aim at the enemy, but it would seem that inevitably, as the war went on and the enemy shot more of one's friends, soldiers became "of value." At times, the Civil War battlefield was a place of roaring artillery, smoke and bullets, the sudden dismemberment of friends and officers: it was the wrath of man rising to a seemingly sublime level. It was the rage of Achilles made visible.

749

The square miles of the eleven Confederate States add up to 749 million. If you add Missouri and West Virginia, which the Union Army had to win and control, but leave out Kentucky, you get 843,000,000.

Some time ago, the present writer was drafted to talk to his son's third-grade class about the Civil War. How does one make clear, and put into proper perspective, the issue of Northern preponderance in population, manufacturing capacity, and infrastructure? How does one do justice to the exhilarating issue of relative capital assets? The students were asked who would win if an NBA star were to play against the smallest girl in the class. The immediate consensus was that the NBA star would win; but one or two wanted to know something first: "What's the game?" What an excellent question, for the game was "who can crawl faster through an 18-inch culvert."

In the Civil War, the game was not to put David and Goliath into a field or a ring and let them slug it out. The game was whether the United States could subdue and retain a hostile territory of approximately 800,000,000 square miles. Napoleon's Europe also contained about 800 million square miles (France 210, Germany 138, Denmark 17, Belgium 12, Luxembourg 1, Austria 32, Italy 116, United Kingdom 94, and Spain 195). Napoleon could not do it; Grant could. Napoleon, in the Duke of Wellington, had a general comparable to Robert E. Lee as chief opponent. The point is that the nature of the Civil War has been misrepresented in a popular notion that pits "overwhelming numbers and resources" against a scrawny, but plucky, underdog.

In facing the United States, the Confederacy encountered a more equal contest than the Colonies encountered against Great Britain during the Revolutionary War. It might have been an uphill battle, but the conclusion of the Civil War was not inevitable. The United States could not simply dump in more numbers and resources; Grant could not simply feed in more men. The great question about the Civil War is not "How did the Confederacy endure so long?" but "Why did the Union prevail?" Northerners were not under siege, not in constant dread of invasion, and their interests were not directly threatened. Why did they fight it out?

In other words, the nature of Civil War combat was to a significant degree a battle of wills rooted in ideology, beliefs, convictions, values—invisibles and intangibles. If nothing else, this means that the Union soldier was probably a more willing and tenacious fighter—a better soldier and a more dangerous enemy—than the popular imagination might picture him.

545

It took courage, discipline, and rage to attack a position. Sometimes it could require all that and more to defend one.

On July 3 at the stone wall at Gettysburg, two Philadelphia regiments made up the thin blue line that faced Pickett's Charge. The 69th Pennsylvania contained 284 men ready for battle; the 71st Pennsylvania counted 261.[22] The total was 545. When the pressing crowd of several thousand yelling, firing, enraged survivors of Pickett and Pettigrew's advance across the open fields reached the wall, would these enormously outnumbered 545 stand or run?

Today, battlefield guides recount in dramatic detail the valorous and harrowing 20 minute advance made by 12,000 Confederates into shot, shell, canister, and sheets of small arms fire—as well they should. What is rare is to hear a description of what it might have looked and felt like to be one of the few soldiers who faced that teeming crowd of armed men. The Philadelphians lost 43 percent of their men (235), only 36 of them listed as missing.

These numbers cannot answer the question of why Civil War soldiers did what they did: they intensify the question. Fighting "for cause and comrades,"[23] Civil War soldiers created the desperate scenes of battle that became a means of defining America. Surely these means must have effects on the end.

5 Percent

As the Confederate army passed through Frederick, Maryland, on September 10, 1862, Dr. Lewis H. Steiner counted them. An inspector for the U.S. Sanitary Commission, a physician, and a professor of chemistry, Dr. Steiner was accustomed to making careful and precise observations. From the alarmist tone of rumors and newspaper reports, he expected Lee to have 100,000 men, but his count yielded only 64,000. Three of Lee's divisions did not pass through Frederick that day. The total figure for the Army of Northern Virginia would have approached 75,000, seven days before the battle outside Sharpsburg, by which time hunger, freebooting, fatigue, and disagreement with the invasion had reduced Lee's army to about 40,000, or perhaps 44,000 by Northern count.

"Northern count"? As is well known, Confederates counted their numbers differently from Federals. The Union army counted everyone present; Lee's figures show only those present for duty in battle. Suppose we take the figures 90,000 Union and 75,000 Confederate at a given battle. The numbers of men actually shooting at each other might be 84,000 Union and 75,000 Confederate. Possibly even more than 6 percent of a Civil War army performed duty other than shooting: cooks, musicians, teamsters, ambulance drivers, drovers.[24] Why did Southern figures not include these noncombatants?

Dr. Steiner supplies what might be one reason. Among the 64,000 men he counted were "over 3,000 negroes." These men were mixed among the army quite thoroughly, dressed in uniforms, and most of them were armed. The doctor did not describe them as shouldering muskets and marching with the infantry, but as filling noncombatant roles: "riding on horses and mules, driving wagons, riding on caissons, in ambulances, with the staff of Generals."[25] "From the Southern point of view, the blacks traveling with the army were servants not soldiers, and they were not included in military formations, deployed for fighting or ever counted in the strength of the army. Yet, in their roles as teamsters, cooks, foragers, nurses, and the like, they freed an equal number of whites from these extra-duty jobs to shoot on the firing line. Scattered references in diaries, letters, and memoirs indicate that privates were as likely to have brought their servants to war as generals."[26]

Little attention has been paid to the role of African-Americans in the Confederate military until recently. The subject is disconcerting from more than one point of view. On one hand, a few blacks actually stood in line of battle apparently voluntarily; on the other hand, thousands of blacks marched along with Southern armies (tens of thousands, as we learn from Segars and Barrow's *Black Southerners in Confederate Armies*[27]), whether or not by choice. Citizens of Greencastle, Pennsylvania, during Lee's second invasion of the North observed that "twenty to thirty slaves" followed each regiment.[28] To attribute to Lee not the reported 40,000 along Antietam Creek, but 44,000, reflects the probability that unlike the whites who fell away, blacks were not given the option of straggling, and to the 3,000 counted by Dr. Steiner must be added a proportionate number in the three divisions that did not pass through Frederick.

The black presence in Southern armies has been familiar in the South. In *Gone with the Wind*, Scarlett sees a column of men coming up one of Atlanta's streets. "There was a great cloud of red dust coming up the street and from the cloud came the sound of the tramping of many feet and a hundred or more negro voices, deep throated, careless, singing a hymn. Rhett pulled the carriage over to the curb, and Scarlett looked curiously at the sweating black men, picks and shovels over their shoulders, shepherded along by an officer and a squad of men wearing the insignia of the engineering corps."[29]

One of the men is a slave from her plantation. She asks whether he has run away. No: "Lawd, Miss Scarlett! Ain' you heerd? Us is ter dig de ditches fer de w'ite gempmums ter hide in w'en de Yankees comes."

The complex relations between whites and blacks is shown by a brief but unforgettable incident in Stark Young's 1934 novel, *So Red the Rose*. The night after Shiloh, one of the servants of a household goes out to the dark battlefield looking for a son of the family: "And he went over the field, feeling all the hair of all the dead till he found Edward, he knew him by his hair. You know how fine it was."[30]

For a Northern soldier, one of the most startling sights of the war would have been coming face to face in battle with an armed black man in Confederate uniform. There were very few such Rebels, and the odds against a close-up encounter must have been great, but in theory such a strange meeting was possible. It would only have increased a Northerner's mystification over Southerners and their cause. The moralism and devotion to the Union exhibited by some Northerners struck Southerners as self-righteous and hypocritical bullying. And while it is a simple enough fact that slavery is wrong and that racism is bad, Northerners reduced the complexity of Southern feelings, interests, and values to those two simplicities. Each side, that is, made caricatures of the other. But why would a black man take up arms against the Union that was trying to free him, and why would white Southerners put rifles into the hands of slaves?—these questions would confuse the caricature beyond recognition. It was bad enough that in both South and North some women— about which gender Victorian Americans had very definite notions—put on uniforms and went into battle.[31] But this master-slave relationship was outside Northern experience and therefore beyond comprehension. Approximately 180,000 black Americans served voluntarily in Union armies, so the relatively few blacks on the firing line under the Stars and Bars do not reverse the consensus of history, but they are inexplicable to the Northern mentality and show that an irreducible element of the war will always remain unexplained. No algorithm will ever be constructed to encapsulate all the variables and behavior of America in 1861–1865.[32] Southerners did not care to understand Yankees and certainly were not going to be told what to do by them. Northerners simply did not understand Southerners, and probably still don't.

2/3

The Grant-Lee question is sometimes put as a thought experiment. Suppose Lee had had Grant's numbers: suppose the situation were reversed. Put Lee at the head of the principle Union army, and Grant in charge of the Army of Richmond or whatever he might have called it. What such reversal experiments often neglect is that you have to reverse all the terms, not merely one. Give Lee 100,000 men, and give him Ambrose Burnside. Give Grant 70,000 men, plus Longstreet, Jackson, and Stuart. Give Lee the task of conquering 800 million square miles, and Grant the task of holding out. Probably the only fair thought experiment is to give Lee 70,000 of his own men (Army of Northern Virginia), and Grant 70,000 of his own (Army of the Tennessee). Who wins? As actually happened in the war, both sides would go at each other with the fury of equals.

But after all, had Lee been a Federal general and Grant a Confederate for three years and upward, Lee probably would have won. The reason is neither the Northern soldier's superiority nor Lee's: the reason would be that the North had 2/3 of the available trained military officers, and from the president on down, the North possessed a higher number of capable leaders. The North had a deeper bench, and as the war went on, the South faced a deepening crisis of command even as the better generals of the Union emerged. If the Union won with "overwhelming numbers and resources," those numbers and resources were its leaders.

Leadership won the Civil War. Both sides had superb common soldiers. The North, with superior numbers and supplies and capital and cannons and rolling stock, could not win in the East under McClellan, Pope, Hooker, and Burnside. In the West, with larger obstacles and no preponderance of numbers, Ulysses S. Grant fairly put the war away.

In *Touched with Fire: The Land War in the South Pacific*, Eric Bergerud writes that his task is "to find the point where the coherence of war meets the brutal experience that confronts those who fight it."[33] That "point" is leadership. A good officer brings formal knowledge and tries to develop and impose coherence—while dealing with the men who encounter the mayhem and chaos of battle, who need supplies, weapons, rest, courage, hope, purpose, discipline, education, and organization. In the officer the textbook and the terror meet. When you have a war of equals, the side with more good leaders wins. The North's great generals might not have been better than the South's great generals, man for man, but there were more of them.

7700

This number—7700—is the figure that Thomas Livermore assigns to Confederate losses in the Wilderness,[34] and the reason for it reveals an imaginary and possibly inaccurate view of combat in the East. No one knows whether or how correct the number is, but the assumption behind it strikes most people as reasonable.

There are figures for only 70 of the 182 regiments of the Army of Northern Virginia at the Wilderness. Even these are probably inaccurate, according to Livermore, because they are based on reports "written long after the battle." Furthermore, after the battle of Gettysburg, some Confederate officers showed that they were not above underreporting their losses if the actual figures reflected badly on themselves or their commands. How, then, might one estimate the CS loss in the Wilderness?

Livermore observes that the Confederate army lost 187 men per thousand a year earlier in "the same thickets" at Chancellorsville. After the Wilderness, Ewell reported a loss of 1250 men, or 68 in each 1000. This is "not credible." Then, Livermore concludes, it is "not extravagant" to assume that Lee's losses were equal to Grant's losses per 1000. That figure was 127.

This sounds reasonable: equal losses, per 1000. The battle was essentially a draw, and Livermore's assumption seems to credit Grant's army with an effectiveness equal to Lee's. But it does not. What seems equal is not equal.

Grant brought 102,000 effectives to the Wilderness, while Lee brought only 61,000, again according to Livermore. But many of those Union troops were not engaged, or were briefly engaged. Assuming a kind of fighting equality or parity between the Union veteran soldier (of whom Grant did not have anywhere near 102,000) and the Confederate veteran, one might think that an equal per-100 rate is reasonable. But there is another way to calculate equal losses: equality of raw numbers. That is, rather than assume that Confederate losses were proportional to Federal losses, why not assume that they were outright equal? In what battle were each army's losses proportional to the other's? Though Confederate numbers were smaller at the Seven Days and Gettysburg battles, their losses were proportionally higher, by a considerable figure, and they were higher outright (28,000 CS and 22,000 US at Gettysburg). Proportionately, Lee lost more than Hooker at Chancellorsville. Why not assume that two days of close combat in the same Wilderness, each side attacking and defending almost equally, would produce the same number of dead, wounded, and captured on each side?

Not to make this kind of assumption is to perpetuate the popular idea that the Yankee was a relatively ineffective soldier. But ultimately, such an assumption reflects badly on both sides. If Lee always faced an army of chumps, his achievement, and that of his men, would seem not so great. But a veteran of Lee's army probably understood all of this when he told Douglas Southall Freeman, "My son, never disparage the Army of the Potomac; it was the greatest army of the age, with the exception of one that modesty forbids me to mention."[35]

65,000–100,000

Lee began the long campaign against Grant with approximately 65,000 men. It is commonly pointed out, with truth, that before the summer was over, Grant lost as many men as Lee had started with, or even somewhat more. Obviously, Lee was the better general and/or the Army of Northern Virginia was the better army.

Or not. Darrell Huff's *How to Lie with Statistics* reminds us that, except in this essay, numbers are no longer pure once they are used in argument: they can be manipulated almost beyond recognition.* The two figures heading this section provide a good example, if not of how to lie with numbers, at least of how to reach opposite conclusions from the same data. Suppose that Lee had lost 100,000 men, as many as Grant had started with. Then one might say, "Obviously Grant was the better general and/or the Army of the Potomac was the better army."

Both numerical statements are correct. Grant's casualties amounted to 56,000 in major battles; add smaller engagements and steady "wastage," as it was called in World War I, and you get between 65,000 and 70,000. Now let us look at Lee's losses. We cannot arrive at them by adding up battle casualties, however; as previously noted, these are inaccurate, incomplete, and/or simply missing.

According to D. S. Freeman, a writer most favorable to Lee and his army, the Army of Northern Virginia counted 59,000 men on June 18, 1864, having received replacements and reinforcements after heavy losses at the Wilderness and Spotsylvania.[36] On July 11, he had 55,000.[37] On September 26, he had 65,000.[38] Freeman says that between May 4 and June 18, Lee had lost "about 30,000" men—probably a considerable understatement, if the considerations made above are correct. But if one accepts Freeman's numbers, Lee had the use of 95,000 men during the summer of 1864, plus non-combatant personnel. But the figure is undoubtedly higher, at least the number Grant started with. And how many of these roughly 100,000 did Lee lose? All of them.

In a military sense, all of those men became a dead loss on or before—and most before—April 9, 1865. We could argue in a football sense, or a playground sense, that one general was better; that is, Lee lost fewer men killed and wounded than Grant. But this is only probably and not certainly true; and in any case such considerations seem not to be pertinent. An army exists to further the objectives of its state, not to win future arguments as to whose men were faster or tougher or wore sharper uniforms.

In question is the relative merit of Grant and Lee, which seems like

---

* The author of this study has not personally read Huff's book nor applied its techniques to his work.—Editor's note.

a pointless exercise in debate or a stale conflict of sectional pride—but it is really a key issue in understanding the nature and consequences of battle in the Civil War. That Grant versus Lee is still debated indicates two things: a) In fact the soldiery was evenly matched at ground level; i.e., we could not go on with this seemingly endless argument if it were clear that either Southern or Northern soldiers were superior. Were one group better than the other, it would be easy to see which general performed better—either because he did well with inferior troops, or because he could not force a decisive issue with superior ones. The most exhaustive study of Grant versus Lee is the series on the Overland Campaign written by Gordon Rhea, who concludes, "Grant and Lee were about as evenly matched in military talent as any two opposing generals have ever been."[39] Grant's was the larger army, with the larger task, possessing superior supplies and dependent upon a more problematic supply line, having a great many men and relatively few experienced veterans, burdened by a recalcitrant command structure—and the result, according to Rhea, was a "stalemate" in the summer of 1864. What we should conclude from all this is that Civil War combat was a fight between equals, the Northern and Southern soldier, and two such armies would not produce a battle of annihilation. The war would not be won by a brilliant battle that dashed the enemy army to pieces. This reality has been held against Grant since 1864, but his accurate perception of Civil War combat induced him to fight the only kind of campaign that could win. His campaign is notable for flank maneuver, not for frontal assault, and it ended successfully, as was not the case with previous Union commanders, who went in search of the smashing victory. Lee himself thought in these terms, as his statement after the Seven Days indicated, to the effect that under normal circumstances the enemy would have been destroyed. Civil War armies did not destroy other Civil War armies.[40] That the Civil War was a fight between equals should have been obvious. Why would Americans, of all people, not be equal to each other?

But—and here is b), which bears on consequences: this is the very thing at issue, even after 140 years. We are still claiming that America was not a single entity, but instead was an uneasy partnership consisting of at least two groups. The two groups, Northern and Southern, were different from each other, more different than alike. And the difference in quality was shown by who could fight better. A couple of primary school kids putting up their dukes on a playground could not have put it more satisfactorily. If the Civil War showed one thing, it was that "all men are cre-

ated equal," just as the Founders declared on July 4, 1776; more particularly, all Americans are equal. Northern soldiers and Southern soldiers,
given proper training, equipment, and leadership, were equal, and left to
themselves would fight each other to a nubs. Hence the 6,500,000 dead in
today's terms—a number that pummels the imagination and makes you
ask "Why?"

One reason why is shown by the continuing Grant and Lee issue. Grant
is a surrogate for Northerners, and Lee represents Southerners, like the
Michigan Wolverines and the Crimson Tide. If Grant was an inferior
general, then the North won only with overwhelming numbers, proving
that Northerners were and still are inferior to Southerners. Honor was at
issue. Southerners fought for honor in 1861, and many still insist upon it
now. Perhaps honor is an unacknowledged reason for Northern persis-
tence as well.

Americans do not get fired up today over who was better: Scipio or
Hannibal, Napoleon or Wellington, Rommel or Montgomery. Each had
his virtues and flaws, military and personal. But the Grant or Lee question still seems to be alive, still arouses emotion and partisanship and inflammatory T-shirts—as if the primary question about the Civil War is
whether it was a fair fight. Neither Grant, Lee, or their men would have
been pleased by such a self-serving reduction and trivialization of their
blood, sweat, and tears. Do we want our great-grandchildren to argue
about whether we, or the terrorists of 9/11, were tougher; or do we want
them to live in peace, with liberty and justice for all? The Civil War turned
out to be about what and who America is. The rest of the world seems to
get it: before he blows you up, a terrorist doesn't inquire as to whether you
are a Northerner or a Southerner. To him, all that matters is whether or
not you're an American. In terms of the Civil War, 150 years later, maybe
that's all we need to know, too.

o

This is the number on a scale of 1–10 that indicates how well we can pre-
dict the future. But understanding that the Civil War was an overwhelming crisis, of unrelenting tense uncertainty, implies that a profound stamp
or habit has been laid upon us, the children of that horror. That catastrophic experience must have given America, or confirmed in America, a
virtually indelible personality as a nation.

We are a people who solved a fundamental political and moral dispute with violence. That is significant and it has marked the American character. But it must also be said that our predecessors fought for liberty. That they died in unthinkable numbers has been seen as meaningful or excusable only because a great moral question was resolved.

We pay heavily when we forget who we are. We are a violent, freedom-loving people to whom right and wrong matter.

Notes

1. Robert Famighetti, *The World Almanac & Book of Facts 1997* (Mahwah, N.J.: World Almanac Books, 1996), 380.

2. See James M. McPherson, *Drawn with the Sword: Reflections on the American Civil War* (New York: Oxford University Press, 1996), 66.

3. Charles Teague, *Gettysburg by the Numbers* (Gettysburg, Pa.: Adams County Historical Society, 2006), 42.

4. Alan Gaff, *Brave Men's Tears* (Dayton, Ohio: Morningside, 1988), 158–161.

5. John W. Busey and David G. Martin, *Regimental Strengths and Losses at Gettysburg* (Hightstown, N.J.: Longstreet House, 1982), 260; John W. Busey, *These Honored Dead: The Union Casualties at Gettysburg* (Hightstown, N.J.: Longstreet House, 1988), 67.

6. Busey and Martin, *Regimental Strengths*, 67.

7. Paddy Griffith, *Battle Tactics of the Civil War* (New Haven, Conn.: Yale University Press, 1987), 139–140.

8. Dave Grossman, *On Killing: The Psychological Cost of Learning to Kill in War and Society* (Boston: Little, Brown and Company, 1995), 24.

9. Grossman, *On Killing*, 250.

10. Teague, *Gettysburg by the Numbers*, 59.

11. Grossman, *On Killing*, 250.

12. Busey and Martin, *Regimental Strengths*, 243.

13. Harry W. Pfanz, *Gettysburg—Culp's Hill and Cemetery Hill* (Chapel Hill: University of North Carolina Press, 1993), 274.

14. Griffith, *Battle Tactics*, 147 and 149.

15. Griffith, *Battle Tactics*, 87.

16. Thomas L. Livermore, *Numbers and Losses in the Civil War in America 1861–65* (Dayton, Ohio: Morningside, 1986), 96.

17. Quoted in Stephen W. Sears, *Landscape Turned Red: The Battle of Antietam* (New Haven, Conn.: Ticknor & Fields, 1983), 209.

18. Homer, *The Iliad*, translated by Robert Fagles (New York: Penguin, 1990), 77.

19. Rick Atkinson, *An Army at Dawn: The War in North Africa, 1942–1943* (New York: Henry Holt and Company, 2002), 46.

20. Atkinson, *Army at Dawn*, 462.

21. Clarence Clough Buell and Robert Underwood Johnson, *Battles and Leaders of the Civil War*, vol. 4 (New York: Thomas Yoseloff, 1956), 176. Another officer reported (p. 177) that an "oak-tree, twenty inches in diameter, was cut down by bullets." This is evidence of rage and of the inability to shoot accurately—unless it is to be interpreted as meaning the soldiers preferred shooting at the tree to taking human life.

22. Busey and Martin, *Regimental Strengths*, 34.

23. The quoted phrase is the title summarizing the thesis of James M. McPherson's book on soldier motivation in the Civil War, *For Cause and Comrades: Why Men Fought in the Civil War* (New York: Oxford University Press, 1997).

24. Lee's army had collected approximately 26,000 head of cattle and 22,000 sheep in Maryland and Pennsylvania in June, 1863. Ted Alexander and W. P. Conrad, *When War Passed this Way* (Greencastle, Pa.: Greencastle Bicentennial Publication, 1982), 140.

25. Quoted in Joseph P. Harsh, *Taken at the Flood: Robert E. Lee and Confederate Strategy in the Maryland Campaign of 1862* (Kent, Ohio: The Kent State University Press, 1999), 169.

26. Ibid., 170.

27. J. H. Segars and Charles Kelly Barrow, eds., *Black Southerners in Confederate Armies* (Gretna, La.: Pelican Publishing Co., 2001).

28. Alexander and Conrad, *When War Passed*, 167.

29. Margaret Mitchell, *Gone With the Wind* (New York: Macmillan, 1936), 306.

30. Stark Young, *So Red the Rose* (New York: C. Scribner's Sons, 1934; repr., Nashville, Tenn.: J. S. Sanders & Company, 1992), 202. Citation is to the 1992 edition.

31. See James Janega, "Favorite Son a Woman," *Chicago Tribune*, November 12, 2006.

32. An algorithm is a recursive computational procedure guaranteeing an accurate solution. It is also a mathematical term for a tune about climate change.

33. Eric Bergerud, *Touched with Fire: The Land War in the South Pacific* (New York: Penguin, 1996), xiii.

34. Livermore, *Numbers and Losses*, 111.

35. Douglas Southall Freeman, *Lee's Lieutenants: Cedar Mountain to Chancellorsville* (New York: Charles Scribner's Sons, 1943), xiii–xiv.

36. *R. E. Lee: A Biography*, vol. 4 (New York: Charles Scribner's Sons, 1935), 444.

37. Ibid., 469.

38. Ibid., 497.

39. Gordon Rhea, *Cold Harbor: Grant and Lee May 26—June 3, 1864* (Baton Rouge: Louisiana State University Press, 2002), xiii.

40. It might be objected that Thomas did this to Hood; but it could be said that Hood did it to himself. In any case, Hood's army was not on a par with Thomas's at that late point in the war.

Afterword

Kent Gramm

A chief consequence of the Civil War is the concept we have of it today. If our concept of what the war meant encourages us to do good and to make our own troubled corner of history a better place, then it continues to work for the best hopes of humankind. But if it makes us more settled into our hatreds and fears, then it may again be said that "the evil men do lives after them." We are the consequences of the Civil War.

As is always the case, the participants of history have put the results of their actions into our hands. If we owe the Civil War generation anything more than gratitude, it is honesty; it is diligence in probing for the truth; and, following their example, we also owe them the decency of acting faithfully to the truth. If we do not owe these things to the dead, we at least owe them to the generations that follow us. But even after 145 years, we are still uncertain of what the nature and consequences of the Civil War were. A great deal of work needs to be done.

No study of the Civil War exists as an equivalent of Paul Fussell's study of what the First World War did to the minds of Europeans (*The Great War and Modern Memory*). Many of the horrors he reports as having been introduced to Europeans in that war were actually experienced by Americans fifty years earlier. But it seems that the Civil War did not have the same stunning effect on Americans that the World War had on Europeans, though the outrages were similar and the losses were even worse. Is that because Americans exercised a general denial such as Alan Nolan suggests was exercised in the specific area of race and politics; or is the difference due to the clearer purposes of each cause in the Civil War, along with the sentimentality and religiosity of each side? Is there something particularly American that the contrast might bring to light? When one in four men of military age in the South died, and when the country as a whole lost today's equivalent of more than six million dead, there must be consequences for the way a country thinks, acts, and believes. No study of the Civil War matches the studies of effects of plague, famine, climate change, and war on fourteenth-century Europeans—Huizinga's *The Waning of the Middle Ages* and Tuchman's *A Distant Mirror*. Such work would be timely, because it is becoming increasingly clear that we are facing the Four Horsemen ourselves.

The Civil War did not teach us that secession was wrong, or that attempts to enslave another race will always fail, or even that right makes might: the Civil War taught us that a democracy *can* defend itself. That is what Abraham Lincoln's "last, best hope" comes down to: it *can* be done; it is possible to defend a democracy. Today, when we are uncertain of what we are defending and confused as to what we will have to surrender in order to protect ourselves, the Civil War is a lesson worth studying.

A study of the sufferings of human beings makes us moralists, whether we wish it or not. But does it make us moral? This is a book of horrors. At the Symposium, there were no screams, no groans, nobody fainted, nobody became ill when Scott Hartwig described artillery projectiles and what they were used for, or when Bruce Evans described the damage a .58-caliber lead slug does to the human body, or when Eric Dean described how men lost their minds after having to maim and kill other men, or when Alan Nolan described the post-Reconstruction denial of what the war could have accomplished.

Already during the war itself, Mathew Brady and Alexander Gardner brought its horrors into American living rooms with their photographs of the dead at Antietam and Gettysburg. You would think that the public would have screamed and groaned and cried "Enough!" But the war continued. A few years ago *The New York Review of Books* carried an article called "Archives of Horror"[1] about books of photographs recording atrocities in the twentieth century. It also concludes that the world has not been changed by such images. Perhaps such photographs, and such symposia as the one upon which this book is based, simply inure us to gruesomeness and inhumanity. Perhaps they are merely ways of coping with the unthinkable.

We have described madness; we have described evil. War is not "all glory," General Sherman said; "it is all hell." The appropriate environment for this subject is not a serene and beautiful national military park, but Dante's Inferno. What are the nature and effects of madness? What are the consequences of evil? Must we abandon all hope when we enter such study? A generation ago, roughly during the time of the Civil War Centennnial, when America was beginning to be divided by another war, a song asked, "How many deaths will it take 'till we know that too many people have died?" The answers to these questions are not found in History.

Note

1. By Charles Simic. *The New York Review of Books,* vol. 50, number 7 (May 1, 2003), 8–11.

Index